Invitation to
Physical Chemistry

Invitation to
Physical Chemistry

Gopala Krishna Vemulapalli
University of Arizona, USA

Imperial College Press

Published by

Imperial College Press
57 Shelton Street
Covent Garden
London WC2H 9HE

Distributed by

World Scientific Publishing Co. Pte. Ltd.
5 Toh Tuck Link, Singapore 596224
USA office: 27 Warren Street, Suite 401-402, Hackensack, NJ 07601
UK office: 57 Shelton Street, Covent Garden, London WC2H 9HE

British Library Cataloguing-in-Publication Data
A catalogue record for this book is available from the British Library.

INVITATION TO PHYSICAL CHEMISTRY
(With CD-ROM)

ISBN-13 978-1-84816-301-0
ISBN-10 1-84816-301-0

Typeset by Stallion Press
Email: enquiries@stallionpress.com

Printed by FuIsland Offset Printing (S) Pte Ltd. Singapore

For Carolyn

Acknowledgments

I thank Dr. Krishna Muralidharan for his help with figures and Professor Steven Kukolich for his continued encouragement. Professor Arvind Marathay has often suggested the need for a book like this one and gave me the momentum to go on. I thank him for his support. For many years, Professor Henry Byerly has had a marked influence on the way I started thinking about science and how it should be taught. I owe him my gratitude.

I am fortunate to have Ms Ling Xiao as the editor. I thank her for many hours she has devoted to this project and suggested improvements. With a background in chemistry, Carolyn, my wife, read the manuscript several times, improving the style and substance, banishing vagueness and curtailing prolixity. If you find the book still lacking in that department, I am the source, since I cannot resist adding and subtracting even after the manuscript is ready to go. Thank you, Carolyn, for the hours you spent on these pages.

Preface

This short book is primarily intended to be read. If that sounds somewhat redundant, think of the present state of textbooks in science. They have become obese to the point of causing health problems to those students who take the trouble to carry them in their backpacks or read them in bed while drifting into the land of Nod. These massive tomes are often referred to, but rarely read. However, on the positive side, the current crop of textbooks in physical chemistry contain a wealth of information, numerous examples and a thousand or more problems, all carefully assembled. Yet one gets the uneasy feeling, while turning the pages, that the sum of that information does not equal knowledge. If knowledge is structured information, a constant flood of information leads to the question Eliot raised: where is the knowledge lost in the information?

My aims in writing this book are: to draw your attention to the basics, to show the warp and weft of physical chemistry and to be comprehensive and concise in explaining the fundamental ideas. This is an invitation to physical chemistry, but it is not a watered-down version and it does not conceal difficult concepts with a patina of facile analogies.

There are at least two ways of studying a subject. One is to stop at the end of each step, go through several examples and set a number of problems before moving on to the next step. The other is to read several connected chapters together before working through the details. Each method has its merits and its disadvantages. The textbooks already on the shelf are meant for the first mode of study. This book is intended to follow the second. I have, therefore, put examples, exercises and the intermediate steps in the derivations on a disc for later study. It is my hope that you will read several chapters once or twice, seated comfortably with a pencil in hand, before even looking at the disc's contents. Our understanding of the topic depends on the context. Words that might initially look strange and foreign start making sense as they fit into the ongoing explanation.

For each chapter in the book the disc includes a) examples, exercises and comments on the derivations, b) additional material that is a natural extension of the text and c) historical and philosophical comments. I resisted the temptation to make it encyclopedic simply because so much useful information is already available in various books and postings on the web by authorities and amateurs. What is needed is coordination, and it is my hope that this book will facilitate that.

I believe that this book can be useful on several levels. If you have already taken a course or two in physical chemistry, you will find a refreshingly unique viewpoint in each section of the book. The way that quantum theory, thermodynamics and statistical thermodynamics are introduced differs significantly from the existing literature.

If you are a student about to take a course in physical chemistry intended for chemists, hyphenated chemists or engineers, willingly or unwillingly, you will gain some idea of the forest before counting the trees and twigs. My method of not stopping at every point and giving a lengthy explanation might be puzzling at first, but as you keep reading you will see patterns emerging, and ideas sticking together and making sense. Don't overestimate the difficulty and don't underestimate the patience needed.

If you happen to be an instructor using this book in a course (how delighted I would be!), I believe the text and the disc will provide material on the same level as the current standard textbooks.

However, this book serves best if the students are required to read ahead, then have a group discussion with the instructor or an assistant before diving into the details. Centuries ago Francis Bacon wrote, "Reading maketh a full man, conference a ready man and writing an exact man." I am sure Bacon would white-out "man" and scribble "person" on the parchment, substitute "discussion" for "conference" and say something about word processing making a legible person, should he be writing today. It is through conference and discussion, however brief, that books of this sort find their best use.

On a more personal note, physical chemistry has always fascinated me. For me it is where the rubber meets the road. Physics and mathematics are both subjects I admire and keep studying, but they are a few inches above the road. Organic, inorganic and analytical chemistry, along with material science, have given us many gifts; they are the real road. I, who do not have the abstract talents of a mathematician, or interest in practical applications, found a comfortable home in physical chemistry. I hope this book conveys some of my fascination with the subject.

If you are broadly interested in science, I urge you to read either this or some other book on physical chemistry. Our knowledge about moon to Mars and beyond, and about mantle and crust, and so forth, comes from spectroscopy, thermodynamics and kinetics. Those are the ideas that have permeated other areas of science through physical chemistry, and the reward of being acquainted with this subject is a deeper understanding of the rest of science.

G. K. Vemulapalli
Tucson, Arizona
June 2009

Contents

Part A

Quantum Chemistry

A revolution took place in science when Werner Heisenberg (1925) and Erwin Schrödinger (1926) independently developed quantum theory as we use it today. There is no major area of science left today that is unaffected by this theory. Explanations for many puzzling phenomena in physics and chemistry fell out (literally) from the theory within a year or two of its formulation. Among them are the nature of the chemical bond, the mechanism for radioactive decay and the connection between structure and spectroscopy. It is not an exaggeration to say that no other theory has achieved such prominence in so short a time. Quantum theory has also been an engine for the development of new technologies, such as the electron microscope and magnetic resonance imaging.

You will find the study of quantum theory both fascinating and rewarding.

Chapter 1

The Schrödinger Equation, Waves and Wave Packets

1.1. Critical Experiment

The experiment that I will describe was among many such tests which directly lead to the heart of quantum theory. You may be familiar with Young's double-slit experiment, which shows that light waves interfere to form dark and bright bands. This experiment dispelled any lingering doubts about the wave nature of light since waves, and only waves, can show such behavior.

What happens when you use electrons instead of light? These experiments were done By Dr. A. Tonomura and associates in Advanced Research Laboratories of Hitachi. Figure 1.1 shows the results of a decisive experiment. Notice the interference pattern projected on the screen. It may seem surprising that electrons show the characteristic interference pattern of waves, but there is a bigger surprise to come. In this experiment, an electron goes through the slit area before the next is ejected from the sharp tip of the electrode. There is no way of avoiding the conclusion that *each* electron forms a complete wave and goes through both slits; otherwise we would not see the diffraction pattern in Fig. 1.1.

If the wave pattern were due to the collective movement of many particles, as is the case with water waves, no one would give a second

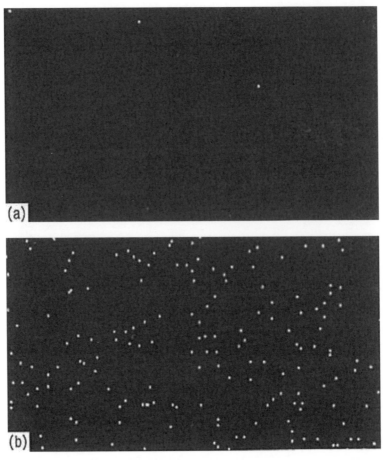

Fig. 1.1. A field emission electron microscope with a biprism is used to record the observed interference patterns in the figure. The biprism functions in the same way as two slits in the optical experiments. This experiment is unique in that one electron wave packet at a time goes through the biprism plane. The frames in the figure show the emergence of the electron interference pattern. The numbers of electrons recorded in the five frames shown are: (a) 5, (b) 200, (c) 6,000, (d) 40,000, and (e) 140,000 (reproduced with the kind permission of Dr. Akira Tonomura, Hitachi, Ltd., Japan).

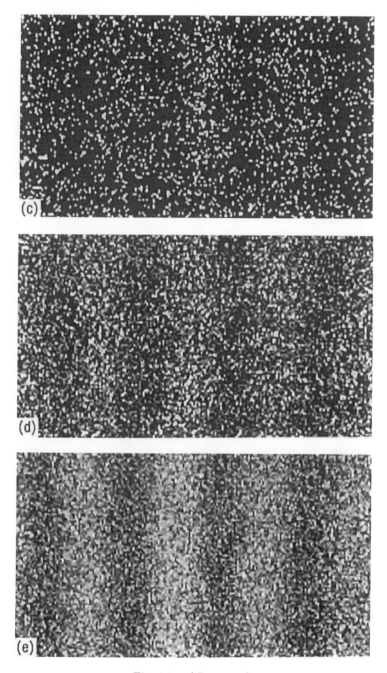

Fig. 1.1.   (*Continued*)

thought to this experiment. But that is not what is happening. It is worth repeating: *each electron shows wave-like behavior*; it apparently goes through both slits simultaneously like a wave.

If we look at the series of photographs in Fig. 1.1 we get another surprise. Each electron makes a dot on the screen just like a particle. These results are so puzzling and so unexpected that this and similar types of experiments have been done many times, under widely different conditions, not just with electrons but also with neutrons, protons, hydrogen molecules, helium and heavier atoms. Similar results have been obtained. Each case begs the same question: How could objects we assume to be particles display wave-like behavior?

1.2. Wave Packets

Puzzling as this behavior is, it can be satisfactorily explained if we accept the following postulate with one essential proviso to be added later:

Elementary units of matter and radiation are wave packets.

A wave packet is a *superposition* (summation or integration) of a very large number of waves of different frequencies. Figure 1.2 gives the profile of a wave packet generated by adding several waves. Leaving the details for a later time, we can say this much now: since it is a collection of waves, a wave packet shows interference when passing through the slits; since it is compact, a wave packet hits the screen like a particle when there is no slit. What we are seeing in this experiment is the effect of change in the size of the wave packet as it passes from the tip of the electrode to the screen.

Think of the electron before it is ejected as a concentrated wave packet confined in a solid material. That wave packet, liberated from the tip, spreads out. By the time it reaches the slits it is large enough to cover the whole slit width. At the screen the wave packet condenses again into a smaller unit to appear as though it were a particle on the screen. Why do I accept this view? Spectroscopic experiments show that electrons embedded in solids show the properties of wave packets and not of tiny balls. You may think that an atom in a solid

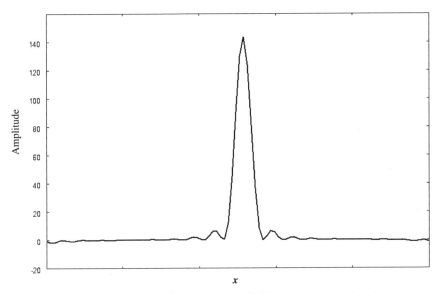

Fig. 1.2. The superposition of many waves of different wavelengths gives a wave packet which exhibits both wave-like and particle-like properties. In this figure, 21 sine waves are added with different weights to form a wave packet. The wave packet behaves like a wave near small objects and shows diffraction. It behaves like a particle near large objects, scattering in near classical trajectories.

state is a particle. Quantum theory says that this is a valid and useful picture under many conditions, but that an atom is never at rest and its properties can *only* be understood by starting with wave theory.

This explanation raises two questions. Firstly, why does the wave packet change its size in travelling from the tip to the detector? Secondly, when a collection of wave packets are detected, each wave packet seems to know its "assigned" position on the screen and forms an interference patterns instead of bunching in one spot. How do they collectively form a pattern while behaving independently? How does each wave packet know where to "collapse"?

The first question can be answered unambiguously. As I will show later in the chapter, the size of the wave packet depends on its interaction with the environment. At present, there are no totally satisfactory answers to the second question and the mechanism for collapse of the wave packet is still being discussed in the physics community. Fortunately, we can make sense of quantum theory as applied to chemistry without answers to the second question.

Now for that essential proviso. The idea of wave packets is not new to quantum theory. Fourier, in the early 19th century showed that heat conduction could be explained by the addition of sine waves. What is new and revolutionary is that quantum wave packets have mass, charge and *spin*; properties we think of as exclusive to particles, i.e. smallest indivisible units. Hence electrons, protons and other elementary wave packets are still referred to as particles; indeed, particle physics is an active area of research. But you can be assured that when particle physicists do their theory, they work with wave packets and not with hard units of matter.

Since light does not have mass equivalent or charge, many prominent physicists dislike describing the photon as a particle, which puts it in the same category as electrons. However, the word coined by G.N. Lewis, a physical chemist, came into common use and cannot be avoided. Alternative words like "quanton" or "q-particle" have been suggested to emphasize the quantum nature, but never entered popular currency. The idea of a particle is so ingrained in our thinking that usage of the term continues. We cannot avoid using the word in the following chapters, since it is part of standard quantum vocabulary. You should, however, keep in mind that I do not mean a hard object of homogeneous composition without internal structure, but a wave packet. To remind you I will occasionally put the word in quotes.

A word of caution before we begin. The quantum revolution took place in the 1920s, during which time brilliant scientists schooled in classical mechanics tried to explain quantum phenomena (such as the interference described above) with classical vocabulary. It is, therefore, understandable that they were puzzled by quantum theory and came up with a thoroughly misleading interpretation of the above experiment. Unfortunately, their misleading statements still persist and are repeated in some textbooks and popular literature, causing unnecessary difficulties to students. The following comments should help you avoid some of these booby traps of misunderstanding.

The interference phenomenon described above is said to illustrate "wave–particle duality," a concept that conjures up a profound dilemma. The electron is conceived to be both a particle and a wave at the same time. Because of the mysterious constraints of quantum theory on what can be "observed" (measured), you can never identify the slit through which the electron emerges. B. Hoffman suggested that the duality is beyond physicists' understanding; they could only believe in particles on Monday, Wednesday and Friday, in waves on Tuesday, Thursday and Saturday, and simply pray on Sunday.

However good the joke, it paints a false picture. There is no wave–particle duality, in the sense of the same "thing" having contradictory properties at the same time, but only wave packet continuity throughout the experiment, even if the position at which the wave packet collapses on the screen is statistical in nature.

It is also sometimes said that the electron shows its particle-like behavior by going through one slit, when the other is blocked. This is a misleading statement. What happens in reality (as shown by several experiments) is that the wave packet either bounces back or goes through the open slit demonstrating the diffraction pattern of waves. The first experiment to prove this was done in 1905 by G.I. Taylor. His photograph of the eye of a needle with a very feeble light (no more than one wave packet at a time) showed diffraction fringes.

It might be argued that we cannot know what constitutes matter. Matter could be a collection of yet to be discovered particles, super strings, crawling microscopic worms or something beyond our imagination. But that does not affect my argument that within the confines of the current quantum theory, only waves and wave packets are required to construct models of matter and radiation. Particle-like behavior is a consequence of the discreteness of wave packets.

Quantum theory is not an easy subject, but many rewards await students who venture to understand the basic theory. It raises some very challenging questions for both philosophers and theoreticians. The difficulties are only magnified if one obsesses with such fruitless questions as: How can something be both a particle and a wave? How can a particle go through two slits simultaneously?

1.2.1. *A road map*

Here, in a nutshell, is the way quantum theory applies to chemistry and materials.

(1) We start with a set of mathematical functions for waves (i.e. **wave functions**) obtained by solving the appropriate differential equation for the system of interest. Since relevant differential equations for chemical applications have already been solved, you need only a brief introduction to differential equations but a thorough understanding of the properties of the wave functions.

(2) Judicious **superposition** (mixing) of appropriate wave functions gives the theoretical framework for quantum chemistry. A qualitative and quantitative account of bonding, hybridization, molecular geometry, electronic transitions and behavior of molecules in external fields is explained by superposing **a set of basis functions**.

Often qualitative arguments, with a rough sketch of the wave functions, give extremely valuable insight. In a few cases, the theory has been used to give quantitative estimates of molecular properties without recourse to recondite mathematical techniques.

1.3. Wave Functions and Wave Packets: A Tutorial

1.3.1. *Monochromatic waves*

A wave with a single frequency and wavelength is said to be monochromatic. It cannot exist except in our imagination, but science is good at explaining the palpable by using the information obtained from mental constructs. So let us start with one such wave. The following function for electromagnetic waves is obtained by solving Maxwell's differential equation. It represents a wave propagating in the x-direction in time, t.

$$E_z = E_o \sin\left(\frac{2\pi x}{\lambda} - 2\pi\nu t\right) = E_o \sin(kx - \omega t) \qquad (1.1)$$

Figure 1.3 gives a sketch of the wave function.

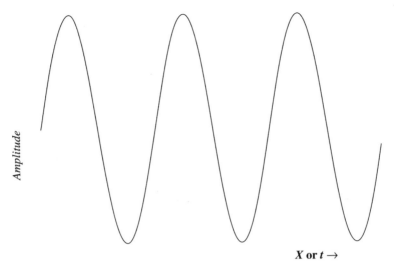

Fig. 1.3. A monochromatic (and monotonous) wave. If the x-axis is distance (X), the distance between crests is the wavelength. If the x-axis is time (t), the distance between crests shows the period of oscillation, the inverse of which is the frequency. It should be noted that the monochromatic wave is theoretical. A wave packet, which is a superposition of such waves, is observed in the experiments.

In Eq. (1.1), E_z is the amplitude of the oscillating electric field and E_o is its maximum value. The rest of the symbols are defined below.

λ: wavelength [m]

$k = 2\pi/\lambda$: wave number [m^{-1}]; it is also called a wave vector when the direction of propagation is considered

ν: frequency of oscillation [s^{-1}]

$\omega = 2\pi\nu$: angular frequency [s^{-1}]

The wave shown in Eq. (1.1) is polarized along the z-direction and *spreads* out along the x-axis. Associated with it is an oscillating magnetic field (not shown) perpendicular to the electric field and the direction of propagation. In a vacuum, frequency and wavelength of electromagnetic radiation are related by the expression

$$\nu\lambda = \frac{\omega}{k} = c \qquad (1.2)$$

where c is the speed of light. Equation (1.1) is also written in the following alternative form:

$$E_z = E_o \sin k(x - ct) \qquad (1.3)$$

1.3.2. *Superposition*

As mentioned earlier, a wave of a single frequency is an idealization. Even the best laser in the world will not emit at a single frequency. Instead, it will exhibit a spread of frequencies. Light from an ordinary lamp has a wide frequency spread. The spread is much, much smaller in laser radiation.

One other point should be kept in mind. In quantum theory we construct wave packets by adding waves as illustrated below. That may lead you to think that waves are the fundamental units of nature and wave packets are "created" by superposition. In reality, matter exists in superposed states, but thinking in terms of individual waves is useful and even inevitable due to the way our human minds function.

With that warning, I will give an example of wave packet construction by adding all waves with k values between $k_o + \delta k$ and $k_o - \delta k$. For illustrative purposes, let us assume each wave between the two limits has the same maximum amplitude, E_o. Then,

$$F(x) = \int_{k_o-\delta k}^{k_o+\delta k} E_o \sin kx \, dk = 2E_o \sin(k_o x) \left[\frac{\sin(\delta kx)}{x} \right] \qquad (1.4)$$

Figure 1.4 gives a plot of the wave packet, function $F(x)$. The figure shows that the wave packet is localized in the neighborhood of $x = 0$. The electric field of each wave with a different k value oscillates between $x = \infty$ and $x = -\infty$. The superposition of these waves, however, localizes the field to a narrow region. The main point to take from this is that as we let k values spread, x values become restricted. The integration in Eq. (1.4) is an example of Fourier transform.

As the size of the wave packet shrinks it appears more and more like a particle. There is some arbitrariness in defining the width of a wave packet. One measure of the width is the distance between the

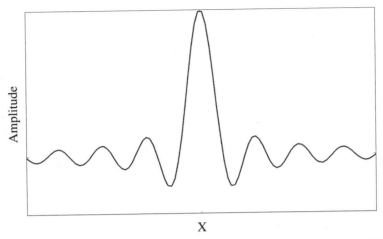

X

Fig. 1.4. Superposition of waves with a phase momentum spread of $\delta k(k = 2\pi/\lambda)$. The width of the prominent peak is taken as a rough measure of the width of the wave packet. According to the Fourier theory $\delta x \cdot \delta k \approx \pi$. When the de Broglie principle is combined with this equation we get the uncertainty principle.

points where the function

$$\left[\frac{\sin(\delta kx)}{x}\right]$$

vanishes for the first time on each side of the central peak (see Fig. 1.4). This function goes to zero when

$$\delta kx_L = -\pi/2 \quad \text{and} \quad \delta kx_R = \pi/2.$$

If we define the width as $\delta x = x_R - x_L$, we have

$$\delta k\, \delta x \approx \pi \qquad (1.5)$$

As the range of waves in the superposition increases, i.e. as δk becomes large, the width of the wave packet decreases.

If we start with the time dependent part of the sine function in Eq. (1.1) and follow the above procedure, we get

$$\delta\nu\, \delta t \approx \pi \qquad (1.6)$$

In this equation, $\delta\nu$ is the range of frequencies in superposition and δt is the duration of beat (for sound waves) or pulse (for light waves) resulting from the addition of frequencies.

We have cut corners in the above discussion. Strictly speaking, Eq. (1.1) cannot be integrated as two separate equations, one as a function of t and another, as a function of x. If k varies, ν will vary simultaneously. Fortunately, Eqs. (1.5) and (1.6) are unaffected by this approximation. The above two equations, which first appeared in Fourier transform theory, form the basis for Heisenberg's uncertainty principle in quantum theory.

1.4. The de Broglie Principle

Scientists are not likely to sacrifice the better parts of their lives doing difficult experiments unless there is a strong theoretical motivation. The impetus for study of "matter waves" came from the following equation, postulated by de Broglie in 1924:

$$p = \frac{h}{\lambda} \tag{1.7}$$

where h is Planck's constant, p is the momentum of the "particle" and λ is its wavelength. If you think of λ as an ever-extending wave and p as the momentum of the smallest object imaginable, you will convince yourself that the de Broglie equation and quantum theory will always remain beyond your comprehension. If, on the other hand, you associate λ with prominent wavelength in the wave packet and p with momentum of that wavelength, you will avoid unnecessary confusion.

There is another way of expressing the de Broglie equation which we will find useful. Since $k = 2\pi/\lambda$, Eq. (1.7) takes the form

$$p = \hbar k \tag{1.8}$$

where $\hbar = h/2\pi$. Kinetic energy, $E(\text{kin})$, is given by

$$E(\text{kin}) = 1/2\, mv^2 = p^2/2m$$

$$(m: \text{mass}; \ v: \text{velocity})$$

Thus we get a relation between kinetic energy, *KE*, and the de Broglie formula:

$$E(\text{kin}) = \frac{h^2}{2m\lambda^2} = \frac{\hbar^2 k^2}{2m} \qquad (1.9)$$

Examples

1. The interference pattern shown in Fig. 1.1 was generated by electrons accelerated to 50 kV. By rearranging the first part of Eq. (1.9) we get the following value for the wavelength of an electron:

$$\lambda = \frac{h}{(2mKE)^{1/2}}$$

$$= \frac{6.6 \times 10^{-34}\,\text{Js}}{[2(9.1 \times 10^{-31}\,\text{kg})(50\,\text{eV})(1.6 \times 10^{-19}\,\text{JeV}^{-1})]^{1/2}} = 0.17\,\text{nm}$$

If the slit separation is much larger than 0.17 nm, we will not observe double-slit interference. Instead we will see single-slit diffraction.

2. Estermann and Stern (1930) studied the diffraction of helium atoms with a speed of 1350 ms^{-1}, using a LiF crystal as the grating. In this case the de Broglie wavelength is 0.07 nm.

$$\lambda = \frac{\hbar}{(mv)} = \frac{6.6 \times 10^{-34}\,\text{Js}}{\left[\left(\dfrac{4 \times 10^{-3}\,\text{kg mol}^{-1}}{6.02 \times 10^{23}\,\text{mol}^{-1}}\right) 1350\,\text{m s}^{-1}\right]} = 0.07\,\text{nm}$$

1.5. Particle in a Box

A ball bouncing around in an empty room is a particle in a box, not an interesting system by any stretch of the imagination. But if you make the room into a box of a few nm^3 volume and the object an electron or an atom, it becomes a very valuable model to study quantum theory. So let us go ahead, setting aside for the moment the relevance of the model to actual physical systems.

Within the box the "particle" feels no force in this model, which is the same as saying its potential energy, *E* (pot), is constant (recall

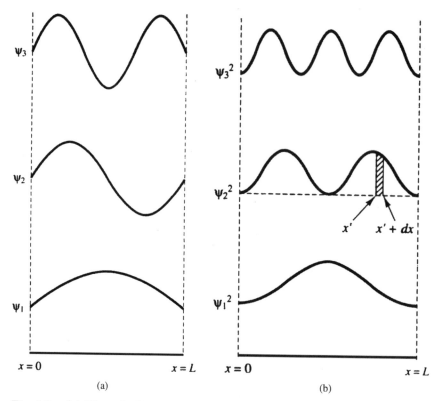

Fig. 1.5. (a) Three de Broglie waves confined between the boundaries at $x = 0$ and $x = L$. The formula for the allowed waves is $\lambda = 2L/n$. (b) Probability density functions (square of the wave functions) for a particle in a box. The shaded area, $\psi_2^2 dx$, is the probability of the particle being in that region.

force $= -[\partial E(\text{pot})/\partial x]$). We can assume $E(\text{pot}) = 0$ without loss of generality since energy, like height, is a relative quantity. We will simplify this system even further by considering motion just in one direction, along the x-axis. A sketch of the box is given in Fig. 1.5(a) where the walls are at $x = 0$ and $x = L$. The potential energy function suddenly becomes infinity at $x = 0$ and $x = L$ or, to put it another way, the particle cannot go past the wall unless it has infinite energy.

The only connection we know between the "particle" and "wave" is that wherever one is the other must also be. They are one and the same. Therefore, this object cannot go past the walls, but because of its wave-like nature, it should be everywhere in between. Only waves that satisfy the equation

$$n\lambda = 2L \qquad (1.10)$$

where $n = 1, 2, 3 \ldots$ meet these criteria. The allowed wavelengths are $\lambda = L/2, L, 3L/2 \ldots$. Since n can only take discrete values, it is called a quantum number.

The substitution of Eq. (1.10) into Eq. (1.9) gives a formula for allowed energy levels:

$$\boxed{E_n(\text{kin}) = \frac{n^2 h^2}{8mL^2}} \qquad (1.11)$$

The energy levels are shown as horizontal lines in Fig. 1.5(a). The wave corresponding to each energy level is also sketched in Fig. 1.5(a).

What have we learned from this exercise? We find that energy is quantized ($n = 1, 2, 3 \ldots$) and that this energy quantization is related to the wave nature of matter. Quantization automatically follows if the waves are confined to a region, i.e. in a potential well of any shape. This is due both to the nature of waves and to the boundary conditions. An appropriate analogy is a string clamped at both ends that vibrates only at certain frequencies, e.g. a violin string.

However, this exercise also leaves us with two puzzles. What are these waves? We can see the waves in Fig. 1.5(a), but not a wave packet, the alter ego of a quantum particle. How can these waves explain wave packet-like behavior? To answer these questions we need the famous Schrödinger equation.

1.6. The Schrödinger Equation

1.6.1. *Development of the equation*

We have in Eq. (1.9) an expression for the kinetic energy of the de Broglie wave packets. Generally speaking wave packets, say electrons in a solid, have both kinetic and potential energy, the latter because of the attraction between the electron and the nuclei. Hence it is natural to ask: what is their total energy? The answer to this question was known nearly 20 years before de Broglie's proposal.

In 1900, Max Planck showed that matter in equilibrium with radiation can exchange energy only in discrete units of $h\nu$, where ν is the frequency of oscillation and h is the constant now associated with his name. Since the universally accepted view until that point had been that the energy of atoms and molecules is a continuous function and that energy exchange with surroundings is unrestricted, and not through discrete units, his discovery marks the beginning of the quantum revolution.

Seven years later Albert Einstein, while developing a theory of heat capacities, came to the conclusion that the *total energy* of each atom in a monatomic crystal changes only in units of $h\nu$. He went on to show that light comes in discrete units, which he called "quanta", each with a total energy of $h\nu$. G.N. Lewis later substituted the name "photons" for Einstein's quanta.

Since the total energy is

$$E(\text{total}) = E(\text{kin}) + E(\text{pot})$$

we get

$$\hbar\omega = \frac{\hbar^2 k^2}{2m} + E(\text{pot}) \qquad (1.12)$$

where the equality $h\nu = \hbar\omega$ and Eq. (1.9) for kinetic energy are used. In wave theory, equations that relate frequency to wavelength (or ω to k) are called **dispersion relations**. They represent the essential physical reality and provide a scaffold for the theory.

Now we are in a position to understand the basis for one of the most famous equations in science, the Schrödinger equation. Many important laws in physical science are formulated as differential equations, since differential equations are very general in their applications. An integrated expression, by contrast, is specific to a particular case. Hence Erwin Schrödinger formulated the differential equation that is consistent with the above dispersion equation. It is deduced from the de Broglie principle and Planck's quantization of energy.

The basic premise is this. A differential equation that gives the above result must have (1) the first derivative with respect to time (since t and ω are a complementary pair) on the left-hand side and (2) the second derivative with respect to x (since x and k are a complementary pair in one dimension) on the right-hand side. Therefore the differential equation that is consistent with Eq. (1.12) has to be

$$-i\hbar \frac{\partial}{\partial t}\psi(x,t) = \left[-\frac{\hbar^2}{2m}\frac{\partial^2}{\partial x^2} + E(\text{pot})\right]\psi(x,t) \qquad (1.13)$$

where $\psi = \psi(x,t)$ is the *wave function* to be found by solving the differential equation. Equation (1.13) is the Schrödinger equation for the one-dimensional case.

The above shows *only* the connection between Eqs. (1.12) and (1.13). Schrödinger could not have convinced the scientific community with such a threadbare argument in 1926. It is not necessary, however, for us to go through the intricate theoretical steps that led to Eq. (1.13).

1.6.2. *Application to the "particle" in a box*

To familiarize ourselves with Eq. (1.13) let us revisit the particle in a box. Since $E(\text{pot}) = 0$ inside the box, the Schrödinger equation is

$$-i\hbar \frac{\partial}{\partial t}\psi(x,t) = -\frac{\hbar^2}{2m}\frac{\partial^2}{\partial x^2}\psi(x,t) \qquad (1.14)$$

This partial differential equation separates into two ordinary differential equations. The wave function in Eq. (1.14) is now a product of two functions, one a function of time and another, a function of x: $\psi(x,t) = \varphi(x)\chi(t)$. Thus we have,

$$-i\hbar \frac{d}{dt}\chi(t) = -E(\text{kin})\chi(t) \qquad (1.15)$$

$$\left[-\frac{\hbar^2}{2m}\frac{d^2}{dx^2}\right]\varphi(x) = E(\text{kin})\varphi(x) \qquad (1.16)$$

The last equation is sometimes written simply as

$$\hat{H}\varphi(x) = E(\text{kin})\varphi(x) \tag{1.17}$$

since it belongs to a class of equations called **eigenvalue equations**. In these, a mathematical **operator** ($\hat{H}$, in the above example) operates on the function to its right and gives back the function unchanged along with a constant. $E(\text{kin})$ is called an **eigenvalue** and φ are **eigenfunctions**. The energy operator, $\hat{H}$, is called the **Hamiltonian** after Hamilton, who showed that in the limit of zero wavelength, physical optics turn into geometric optics. Schrödinger later developed wave mechanics from this idea. We will see a few more examples of eigenvalue equations later.

Going through the standard procedure we find that the solutions for the above equation are

$$\varphi_n = A \sin \frac{n\pi x}{L} \qquad n = 1, 2, 3 \ldots \tag{1.18}$$

$$\chi_n = \exp\left(-\frac{iE_n t}{\hbar}\right) \tag{1.19}$$

and the product function

$$\psi_n = \varphi_n(x)\chi_n(t) = A \sin \frac{n\pi x}{L} e^{-i\frac{E_n}{\hbar}t} \tag{1.20}$$

In these equations, L is the length of the box and A is a constant that has yet to be determined. The allowed energies are given by

$$E_n(\text{kin}) = \frac{n^2\pi^2\hbar^2}{2mL^2} = \frac{n^2 h^2}{8mL^2} \tag{1.21}$$

The results in Eq. (1.21) are the same as those we obtained from the de Broglie principle (Eq. (1.11)) without recourse to a differential equation. What is new and special here?

Let us take a brief look at the history of quantum theory. That atoms absorb and emit light as discrete quanta of energy has puzzled scientists for a long time. In 1913, Bohr developed a theory for the hydrogen atom spectrum, but it relied on questionable assumptions and could not be extended. Schrödinger's theory, on the other hand, has far wider import as we will see. The basics of the quantum theory of matter as it stands now were developed by Schrödinger (differential equations), Heisenberg (matrices) and Dirac (complex algebra) independently and almost simultaneously. Schrödinger's differential equations method, through which we get wave functions, finds the widest use in chemistry. To distinguish the new, powerful theory from Planck's and its extension, it is sometimes called new quantum theory, wave mechanics or matrix mechanics.

Within a few months of the development of the Schrödinger equation, the first quantitative theory of the chemical bond was given by Heitler and London. Many spectacular successes followed within the next few years. Since then, quantum theory has penetrated into every aspect of chemistry and material science. We can now quantitatively estimate bond lengths, bond strengths, dipole moments, quadrupole moments, magnetic properties, spectroscopic frequencies and many other properties starting with quantum postulates. We can also calculate thermodynamic properties from quantum mechanics and statistical thermodynamics. While some actual calculations might require a special mathematical background, many important applications only require a *qualitative understanding* of quantum concepts and wave functions.

The only wave functions introduced so far are those for the particle in a box model. This is all we need now for a tutorial on the properties of wave functions, before we move on to more realistic cases.

1.7. The Uncertainty Principle

Combining the de Broglie equation (Eq. (1.8)) with the result from the Fourier transform (Eq. (1.5)) we get

$$\Delta x \Delta p_x \approx h \qquad (1.22a)$$

Combining equations $E = \hbar \omega$ and Eq. (1.6) we get

$$\Delta E \Delta t \approx h \qquad (1.22b)$$

These are the uncertainty relations as deduced by Heisenberg. The approximation sign is used to indicate that there is some arbitrariness in defining the size of the wave packet. In some books you will find either $\hbar$ or $\hbar/2\pi$ instead of Planck's constant, depending on how uncertainty is defined.

In the very beginning of this chapter we saw that the size of the wave packet changes between generation and detection. The explanation comes from Eq. (1.22a). As the wave packet interacts with the surroundings, its energy and momentum change. The change in momentum is compensated by the corresponding change in size.

It is commonly stated that Δx is the uncertainty in the position of the "particle" and Δp_x, the uncertainty in its momentum. Unfortunately this causes serious confusion for readers who have not been properly introduced to the jargon that scientists fall into. In the actual application of the uncertainty principle, Δx always corresponds to the uncertainty in the *size* of the wave packet and Δp_x, to the uncertainty in **phase momentum** (spread in k-values) which cannot always be equated to classical momentum.

1.8. Interpretation of the Wave Function

It is not necessary for a student of quantum chemistry to understand the variety of techniques used to solve differential equations. It is, however, essential to have a clear understanding of the properties of wave functions and their use in tackling chemical problems.

Particularly important topics are: phase, probability density, calculation of physical properties, phase and group momenta, and recovering the classical idea of particles.

1.8.1. *Phase*

Note that the wave functions for a particle in a box, except for one with the lowest energy, have both positive and negative parts. The terms "positive" and "negative" here refer to relative phases (which determine the interference pattern) of the parts in the wave function, rather than to charge. For functions with imaginary terms, phase is $\pm i$. Phase of wave functions is at the very root of quantum explanations. When you superpose different functions you get interferences because of the phase differences. If we superpose squares of the functions we get no interference, since we are adding only positive functions everywhere. We will see that much of quantum explanation, including the explanation for bonding, depends on interference effects.

1.8.2. *Probability density and probability*

We have a set of functions corresponding to waves, and know that these waves represent energy states. However, the significance of these waves puzzled even Schrödinger. The following interpretation, now universally accepted, was proposed by Max Born.

In electromagnetic fields, the electric field is either positive or negative but the intensity of light, which is obtained by squaring the wave, is either zero or a positive number. If we think of light as a collection of "photons," it makes sense to assume that high intensity indicates a large number of "particles." Based on this analogy, Born suggested that the square of a wave function is related to the particle nature of the object.

The square of the wave function in Eq. (1.20) is

$$\psi^*\psi = A^*A \sin^2 \frac{n\pi x}{L} \exp\left(i\frac{E_n}{\hbar}t\right) \exp\left(-i\frac{E_n}{\hbar}t\right) = |A|^2 \sin^2 \frac{n\pi x}{L} \tag{1.23}$$

(Rule: to square a complex function you multiply it by its complex conjugate, obtained by replacing i by $-i$; $(a + ib)$ and $(a - ib)$ are complex conjugates of one another.) Eq. (1.23) is identical to the square of the wave function in Eq. (1.20), which is independent of time.

We see that the square of a wave function is another function, which is positive everywhere as shown in Fig. 1.5(b). But where is the "particle" to which it corresponds? Born's suggestion was that the square of the wave function gives the probable location of the "particle." According to the Born hypothesis

$$\psi^2(x')dx = \rho(x')dx \tag{1.24}$$

is the probability that the particle is located within the segment dx in the neighborhood of x'.

In the above equation, $\rho(x')$ is called the **probability density** function. It has the dimension length^{-1} for a one-dimensional system. Since there are an infinite number of points between $x = 0$ and $x = L$ (or between $x = 0$ and $x = x'$), it is meaningless to talk about probability at any one point. We can only define probability within the length element dx.

It is understood that a single experiment may locate the particle anywhere between $x=0$ and $x=L$, but a large number of experiments will find it at different positions with the relative weights shown by the square of the function. This is what happens in real experiments. In the experiment we discussed at the beginning of the chapter, each trial localized an electron at a different position. It is the whole set of measurements that provides a map of the function for the wave packet. Another example comes from x-ray diffraction which shows the electron distribution (probability density function) around the nuclei in a molecule. The actual experiment is done on a crystal with a very large number of atoms (and electrons) rather than on one atom. The theoretical values calculated with the above formula correspond quite well with the observed probabilities.

Since total probability of all possible outcomes has to be unity, we require that

$$\int_0^L \psi^*\psi dx = \int_0^L \varphi^2 dx \equiv 1 \qquad (1.25)$$

Functions that satisfy the above conditions are said to be **normalized**. When you evaluate the above integral to determine the constant A, you find that the normalized wave function is given by

$$\varphi_n = \pm\sqrt{\frac{2}{L}}\sin\frac{n\pi x}{L} \qquad (1.26)$$

Example

To illustrate, let us estimate the probability between $x = 0.75L$ and $x = 0.76L$ for a particle in a box.

From the limits given we have

$$dx = 0.01L$$

For the lowest energy function $n = 1$. Hence probability density at $x = 0.75L$ is given by

$$\rho(0.75L) = \left(\frac{2}{L}\right) \sin^2 \left(\frac{0.75L\pi}{L}\right) = \left(\frac{2}{L}\right)\left(\frac{1}{2}\right) = \left(\frac{1}{L}\right)$$

Notice that ρ has the dimension length^{-1}.

$$\text{Probability} = \rho(0.75L)dx = \left(\frac{1}{L}\right)0.01L = 0.01.$$

1.8.3. *Relation to physical properties*

Quantum theory interests chemists because it provides methods for calculating the properties of atoms, molecules and materials. We have seen how we find a set of energies from the Schrödinger equation. We will see later that the energies thus obtained are used for assigning spectral transitions and deducing structure, but that is not the end of the story. Once we have the eigenfunctions we can use them, each alone or in combinations, to calculate many other properties.

In quantum theory, for every physical property, a, there is a corresponding **mathematical operator**. The value of a property, whose operator is, say $\hat{A}$, is calculated by the formula

$$\langle a \rangle = \frac{\int \psi^* \hat{A} \psi dx}{\int \psi^* \psi dx} \tag{1.27}$$

In this formula $\langle a \rangle$ is the **expectation** value of the property whose operator is $\hat{A}$. The denominator goes to unity if we start with normalized functions. For real functions, $\psi^* = \psi$. How this formula came about is easy to see with two examples.

Suppose you want to calculate the average position of a particle in a box. You can see from Figs. 1.5(a) and (b) that it is $L/2$, the midpoint. Suppose we don't know this and want to devise a formula to estimate it. An obvious procedure is to multiply each possible position value by the probability of the particle being at that position, and then to sum all such terms. If the position is a continuous variable, as in the present case, summation becomes integration and probability is replaced by probability density. Hence

$$\langle x \rangle = \int \rho(x) x \, dx \tag{1.28}$$

Since $\rho(x) = \varphi(x)\varphi(x)$, we may write the above expression as

$$\langle x \rangle = \int_o^L \varphi(x)\hat{x}\varphi(x)dx = L/2 \tag{1.29}$$

In this expression, $\hat{x}$ is considered an operator. Its operation consists of multiplying whatever is to the right of it. In this particular case the operator can be placed anywhere within the integral, but that is not valid for all operators.

For example, let us consider the slope of the wave function. We know that slope, S, is given by $(d\psi/dx)$. Hence d/dx is the operator for slope. So the expectation value of the slope is

$$\langle S \rangle = \int_o^L \varphi(x) \left(\frac{\hat{d}}{dx} \right) \varphi(x)dx \tag{1.30}$$

It is a simple exercise to show that $\langle S \rangle = 0$. The important point here is to realize that in quantum usage $\hat{d}/dx$ is called an operator and that it must be placed in the middle in the formula for expectation value. Hence Eq. (1.27) is the general expression for property calculations.

What happens if the function we use is an eigenfunction of the operator? In that case

$$\hat{A}\psi = a\psi \tag{1.31}$$

and a is an eigenvalue. Unlike the position and slope considered above, it is not an average but a definite value. In the formal development of quantum theory, the difference between eigenvalues and expectation values is important, but for the moment we can proceed without paying too much attention to that.

1.8.4. *Phase momentum and group momentum*

Waves have two types of momentum, and it is important to distinguish between them. Consider the wave functions in Eq. (1.20). The sine function oscillates with time, which is readily seen by considering only the real part of the equation

$$\psi_n = \sqrt{\frac{2}{L}} \sin \frac{n\pi x}{L} \cos \frac{E_n t}{\hbar} \tag{1.32}$$

The frequency of oscillation is $\omega = E_n/\hbar$. The momentum corresponding to this oscillation is the **phase momentum**. Phase momentum corresponds to in-place oscillations of the waves. A useful analogy is the sidewinder, a poisonous snake in the American desert whose body undulates perpendicular to its path as it moves forward. The sideways undulations are the manifestation of phase momentum. The forward movement of the snake's body is the group momentum, analogous to a group of waves moving forward. If I see a sidewinder with huge phase momentum and no group momentum, I will stay in place and enjoy the sight. If it has group momentum toward me, I will run. The analogy, however, cannot be stretched too far since phase momentum in quantum theory is not connected to the actual movement of particles.

When you square the function in Eq. (1.20) for a particular n value (keeping in mind the rules for working with imaginary quantities) time drops out and we get stationary probability distribution. How do we then get a particle (as a wave packet) to move? Consider adding (superposing) the first two sine functions in Eq. (1.20) and squaring the sum. The result is the probability density given by

$$\rho(x,t) = \frac{1}{L}\left[\sin^2 \frac{\pi x}{L} + \sin^2 \frac{2\pi x}{L} + 2\sin\frac{\pi x}{L}\sin\frac{2\pi x}{L}\cos\left(\frac{\Delta E t}{\hbar}\right)\right] \tag{1.33}$$

where ΔE is the energy difference between the two states. Figure 1.6 gives a sketch of this probability function. We see from this that the wave packet moves back and forth like a particle with a frequency of $\omega = \Delta E/\hbar$ and corresponding group momentum. Admittedly, this is

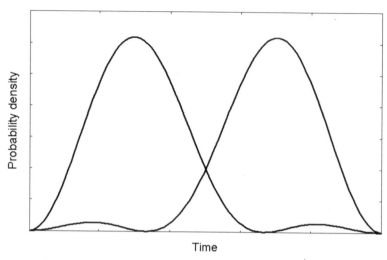

Fig. 1.6. Superposition of two time-dependent functions. The wave packet oscillates with a frequency of $\Delta E/\hbar$, where ΔE is the energy gap between the two functions.

a crude picture of a particle, but consider what happens if we add many more functions. We will get a narrow wave packet which, for all practical purposes, is a particle-like object.

Wave theory gives the following formulas for the two velocities:

$$\text{phase velocity} = \frac{\omega}{k} \tag{1.34}$$

$$\text{group velocity} = \frac{d\omega}{dk} \tag{1.35}$$

From the dispersion equation (Eq. (1.12)) it follows that group velocity (and momentum) is twice the phase velocity (and momentum). However, the concept of group momentum becomes fuzzy if the wave packet's shape and size change with time, which is the case if the wave packet has a large spectrum of k-values.

1.8.5. *Recovering the classical world*

Consider a helium atom in a 10-cm^3 cell. We all agree that it will behave like a particle, bouncing off the walls. Yet if you solve the Schrödinger equation, you get a series of wave functions that cover the whole space in the cell. The square of these wave functions indicates that the probability density is such that the helium atom is simultaneously at every point between the two walls.

How do we resolve this conundrum? According to Fourier theory (Eqs. (1.5) and (1.6)) and Heisenberg's uncertainty principle (Eqs. (1.22a) and (1.22b)), there is actually no puzzle at all. Suppose you have one helium atom in a cell and its energy is somehow fixed to correspond to one of the stationary states. That one atom then occupies the whole length of the cell. Recent interference experiments with single atoms support this conclusion. If that helium atom is one of many in the cell, its energy does not stay constant, which means its k (momentum) spans a wide range. Now the width (size) and momentum variations are connected by the equation $\Delta k \Delta x \approx \pi$. I will omit the details of an order of magnitude estimate, but it emerges that a wave packet of about $\Delta x \approx 10^{-10}$ m (diameter of the He atom) is what we can expect if we take into account energy fluctuations.

Perhaps the most interesting point here is that the *size* of the sub-microscopic object depends on the surroundings, and can be manipulated by controlling its interaction with them. This also applies to the size of the wave packet in the interference experiments I described at the beginning of this chapter.

It would be foolish to extend these arguments to a macroscopic object composed of many particles since the interaction among the particles in the object outweighs the object's interaction with its surroundings. Nevertheless, it does occur in popular parlance, as even the scientists who are masters of their discipline have a tendency to throw caution to the wind when speaking about quantum theory in public.

1.9. Particle in a Three-Dimensional Box

The Hamiltonian (energy operator) for a particle with three degrees of freedom is

$$H = -\frac{\hbar^2}{2m} \left(\frac{\partial^2}{\partial x^2} + \frac{\partial^2}{\partial y^2} + \frac{\partial^2}{\partial z^2} \right) \tag{1.36}$$

If movement of the particle in one direction is independent of movement in the other directions, the energy must be a sum of three terms without any cross terms. In that case the wave function must be a product of three wave functions, each a function of one of the coordinates. This is because the probabilities in each direction must be independent. So without much ado we write

$$\Phi(x, y, z) = \varphi_{n_x} \varphi_{n_y} \varphi_{n_x} = \sqrt{\frac{2}{L_x}} \sqrt{\frac{2}{L_y}} \sqrt{\frac{2}{L_z}} \sin \frac{n\pi x}{L_x} \sin \frac{n\pi y}{L_y} \sin \frac{n\pi z}{L_z}$$

(1.37)

and

$$E = E_x + E_y + E_z = \frac{h^2}{8m} \left(\frac{n_x^2}{L_x^2} + \frac{n_y^2}{L_y^2} + \frac{n_z^2}{L_z^2} \right)$$

(1.38)

Figure 1.7 illustrates a few energy levels for a box with equal sides. The new concept to emerge here is **degeneracy**, which is described in the caption.

1.10. Quantum Wells, Wires and Dots

One of the most interesting aspects of modern science is that some of the quantum theoretical ideas, ideas that appeared stranger than fiction when first suggested, have turned out to be blueprints for revolutionary devices. The electron microscope, based on the de Broglie equation, is one example. Who would have thought before 1924 that one could build a microscope where an electron beam replaces the light source?

The particle in a box is not a theoretical abstraction anymore, even though it was to begin with. By sandwiching a material with low electron energy between two high-energy materials, scientists have been able to confine electrons almost to a plane (well), a line (wire) or a miniscule volume (dot).

Electron energies in gallium arsenide are much lower than those in aluminum gallium arsenide. Hence sandwiching a thin wafer of the former between two layers of the latter creates a two-dimensional box. Strictly speaking, the wafer is not a two-dimensional object,

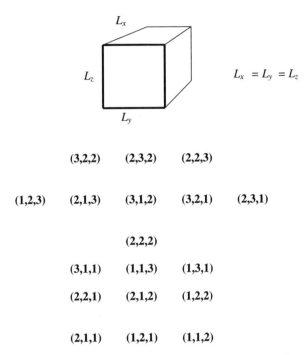

Fig. 1.7.   For a three-dimensional box, the combination of different quantum numbers may lead to the same energy. In the above diagram each bar is an energy *state*. An energy *level* denotes all the states with the same energy. For example, the top level is triply *degenerate* since it has three states with the same energy. There are seven levels and twenty states in the above diagram.

nor is a quantum wire a one-dimensional object since they both have the same thickness. However, as Eq. (1.38) shows, energy increases rapidly with decreasing length. Thus if $L_x > L_y = L_z$, all the low-lying energies will be for movement in the x-direction and the device behaves like a one-dimensional box.

My aim in this chapter has been to introduce you to the new way of thinking required by quantum theory. The important ideas are (1) quantized energies, (2) wave functions, (3) wave-packets and their extension in space (relation between particle-like and wave-like behavior), (4) probability densities, and (5) procedures for calculating molecular properties. We have only considered two examples to illustrate these ideas: a free wave packet (potential energy = 0 everywhere) and a "particle" in a box (potential energy = 0 in the box). You now have sufficient information to explore these topics further.

 Related material on the disc.

With CD-ROM

Chapter 2

Spectroscopy: Rotations and Vibrations

If you were to step back in time to visit chemistry labs of the 1950s, you would see few optical instruments. Possibly there would be a UV-visible spectrometer and a refractometer on the lab bench. Today it would be hard to find a lab without some form of spectrometer and several types of auxiliary apparatus. Spectroscopy has become our eyes into the universe, looking at everything from atoms to stars. It is the technique by which we identify chemicals, deduce molecular structure and gain data needed to compute thermodynamic properties. Its possibilities have not yet been exhausted.

Figure 2.1 shows the portion of the electromagnetic spectrum that concerns us in this chapter and Fig. 2.2 gives an energy level diagram for an idealized diatomic molecule. The energy of such a molecule can be partitioned into four terms:

$$E(\text{total}) = E(\text{electronic}) + E(\text{vibrational})$$

$$+ E(\text{rotational}) + E(\text{translational}) \qquad (2.1)$$

The four modes of motion are coupled to some extent. For instance, rotations stretch the bond length and affect the vibrational frequencies. Hence the above equation is an approximation, but a useful

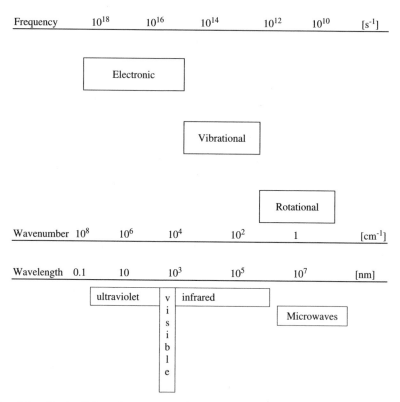

Fig. 2.1. Part of the electromagnetic spectrum that is widely used in spectroscopy. Approximate ranges over which different types of transitions manifest are shown in boxes.

one, as we will see. In this equation the energy terms on the right side are arranged in decreasing order of magnitude. Figure 2.1 also shows ranges over which transitions are usually observed. In this chapter we will only consider rotational and vibrational spectral transitions. Before discussing the actual spectra, however, we need to make ourselves familiar with the quantum theory of rotations and vibrations.

A word on the strategy: The usual way of introducing quantum theory is to start with mechanical models that go back to Galileo, Newton and Hooke. Thus we start with particles confined to boxes, oscillating around a position, rotating around a point, spinning around an axis and so on. Then the model is refined and transformed

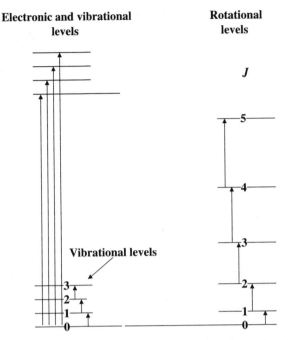

Fig. 2.2. Relative energies of electronic, vibrational and rotational levels. The scale for rotational levels is expanded. There could be many more rotational levels between $v = 0$ and $v = 1$.

into a mathematical model. At this stage we take a deep breath and say that the mechanical model is classical and wrong, but the mathematical model is the real thing. Some scientists have felt uncomfortable with the idea of deriving quantum theory from classical models, only to deny the physical attributes of that model. However, it is the most convenient way to get into quantum theory so that is what we will do.

2.1. The Harmonic Oscillator Model

Galileo's attention, according to one account, was once caught by a swinging chandelier in a cathedral in Pisa. He estimated the time taken for one full oscillation by counting his pulse and found that while the amplitude of the chandelier's oscillation varied with the breeze, the frequency did not. Since then, the **harmonic oscillator** has been the physicist's favorite model, appearing in almost every important theory. Simply defined, a harmonic oscillator is a device that inter-converts potential and kinetic energies at a constant frequency. The most common examples are the simple pendulum and a weight attached to an ideal spring. In his theory of heat capacities of monatomic crystals, Einstein used this model for atoms. We will examine that theory in Chapter 9.

In the harmonic oscillator model, a particle oscillates around its equilibrium position as it experiences a restoring force, f, proportional to displacement, x:

$$f = -kx \qquad (2.2)$$

In this equation, k is called the **force constant**. The negative sign indicates that force opposes displacement.

It seems reasonable to assume that frequency of oscillation, $\nu[s^{-1}]$, depends on the force constant and mass, m, of the oscillator. The dimensions of these quantities are: $k[\text{kg s}^{-2}]$, $m[\text{kg}]$. From matching the dimensions we have

$$\nu[s^{-1}] \propto \sqrt{\frac{k\,[\text{kg s}^{-2}]}{m\,[\text{kg}]}} \qquad (2.3)$$

The actual expression is

$$\boxed{\nu = \frac{1}{2\pi}\sqrt{\frac{k}{m}}} \qquad (2.4)$$

Equation (2.3) is deduced by dimensional analysis which shows the correct form for the mathematical expression, but doesn't always give the numerical constants. However, it leads to an intuitive understanding and visualization of the mathematical relations. Hence dimensional analysis is worth practicing whenever the opportunity arises.

Potential energy, $E(\text{pot})$, of the oscillator is given by

$$E(\text{pot}) = -\int f dx = \int kx dx = \frac{1}{2}kx^2 \qquad (2.5)$$

The integration constant is zero since we can a assign a value zero for potential energy at $x = 0$ without affecting the energy differences

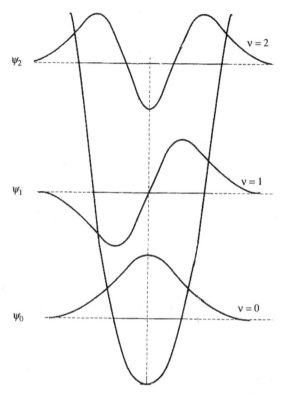

Fig. 2.3. The parabola is the potential energy function for a harmonic oscillator. The horizontal lines represent the energies for 0, 1 and 2 quantum states starting from the bottom. The curves on the horizontal lines are the wave functions for the harmonic oscillator in this example.

needed to understand spectra. Figure 2.3 shows the potential energy function for a harmonic oscillator.

According to classical mechanics, the harmonic oscillator energy varies continuously as displacement increases. This, of course, is not the case in quantum mechanics. We find the quantum energies by solving the Schrödinger equation (Eq. (1.14)) with the potential energy function given in Eq. (2.5). Although the procedure is by no means trivial, the results are quite simple. Quantum energies of the oscillator are given by:

$$E_n = h\nu(\mathrm{v} + 1/2) \tag{2.6}$$

where v is the quantum number, restricted to the values $v = 0, 1, 2, 3 \ldots$ These energies are indicated as horizontal bars in Fig. 2.3. A few eigenfunctions for the harmonic oscillator, obtained from solving the Schrödinger equation, are also plotted in Fig. 2.3.

Here, in summary, is what we learn from quantum theory:

1. Energy is quantized. This is to be expected, since confining waves to a region leads to discrete energy values.
2. The minimum energy (also called **zero-point energy**) for the oscillator is $(1/2)h\nu$. Unlike in the classical theory, the quantum oscillator can never be at rest.
3. The wave functions shown in Fig. 2.3 are similar to those for a particle in a one-dimensional box, except they spill past the potential energy boundary, shown as a parabola in the figure.

So far we have considered the harmonic oscillator theory for a single particle. However, it turns out the theory developed so far also applies to diatomic molecules, if we replace the mass in Eq. (2.4) with the **reduced mass** of the two atoms in the molecule. The reduced mass, m_r, is given by

$$m_r = \frac{m_1 m_2}{m_1 + m_2} \tag{2.7}$$

One further change is needed. We should equate x in Eq. (2.5) to the stretching and compressing of the bond

$$x = r - r_e \tag{2.8}$$

where r is the instantaneous bond distance and r_e, the equilibrium bond length. We can now apply this theory to the vibrational spectra of diatomic molecules.

2.2. The Rigid Rotor

Angular momentum plays as important a role as energy in the quantum theory of atoms, elementary particles and molecular rotors.

Classical mechanics gives the following formula for the rotational kinetic energy of a linear molecule

$$E(\text{rot}) = \frac{L^2}{2I} \qquad (2.9)$$

where L denotes the angular momentum and I, the moment of inertia:

$$I = \sum m_i r_i^2 \qquad (2.10)$$

In this formula, m_i is the mass of the ith particle and r_i is its distance from the center of mass.

For a diatomic molecule, the rigid rotor formula simplifies to

$$I = m_r r_R^2 \qquad (2.11)$$

In this equation m_r is the reduced mass and r_R is the bond length in the rigid rotor approximation.

In quantum mechanics, angular momentum is quantized and the symbol for the angular momentum quantum number is J. The quantum formula for angular momentum is

$$L \text{ values} = \hbar\sqrt{J(J+1)} \quad J = 0, 1, 2 \ldots \qquad (2.12)$$

Substituting this equation into Eq. (2.9) gives the quantized rotational energies for a linear molecule.

$$E_J = \frac{\hbar^2}{2I} J(J+1); \quad J = 0, 1, 2 \ldots \qquad (2.13)$$

$$\equiv BJ(J+1)$$

B is called the rotational constant. Rotational energy levels are also shown in Fig. 2.2. (Equations (2.12) and (2.13) are simultaneously valid only because the operators for energy and angular momentum **commute**. This important property of quantum mechanical operators is best discussed in Section 3.1.3.)

Wavelength in nanometers and wave number in centimeters^{-1} are measured in the infrared region. It is more practical to measure frequency as seconds^{-1} in the microwave region. Hence the rotational constant is often expressed in these units. The symbols for rotational constants in these units are $\tilde{B}[s^{-1}]$ and $\bar{B}[cm^{-1}]$. The formulas for conversion are indicated below

$$\tilde{B}[s^{-1}] = \frac{B[J]}{h} \tag{2.14a}$$

$$\bar{B}[cm^{-1}] = \frac{B[J]}{hc} \tag{2.14b}$$

where c is the speed of light $[cm\,s^{-1}]$. We are now in a position to examine a few spectra.

2.3. Diatomic Molecules

2.3.1. *Hydrogen chloride*

Figure 2.4 gives the infrared gas-phase rotational-vibrational spectrum of $H^{35}Cl$ and $H^{37}Cl$. From a simple spectrum like this we can deduce (a) bond length, (b) force constant, a measure of bond strength, (c) isotopic composition, (d) temperature of the sample and (e) data on stretching of the bond by centrifugal force.

Compared to vibrational energies, rotational energies are closely spaced. Hence we expect many rotational levels between a pair of vibrational levels. At ambient temperature, several rotational levels will be occupied. Hence the vibrational spectrum also consists of transitions between rotational levels. Figure 2.5 shows two vibrational levels ($v = 0$ and $v = 1$) and a few rotational levels in each vibrational level.

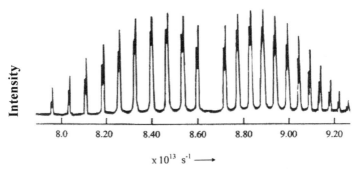

Fig. 2.4. Infrared, rotational-vibrational spectrum of HCl, showing the P branch on the left-hand side.

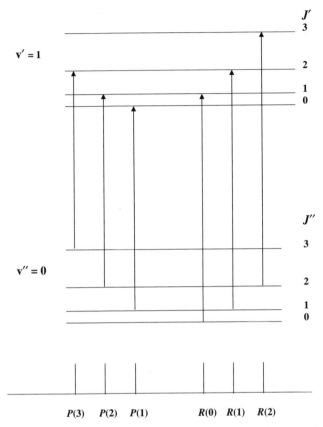

Fig. 2.5. Schematic representation of rotational structure of a vibrational band. Transitions with $\Delta J = -1$ form the P branch. Transitions with $\Delta J = +1$ form the R branch. Expected positions of the rotational transitions are shown at the bottom. Intensities of the rotational transitions depend on the Boltzmann distribution.

Two factors control the appearance of the spectrum. These are (a) the selection rules for allowed transitions and (b) the population of the energy levels. The selection rules are:

(1) The molecule must have a dipole moment or a capacity to acquire one during transition. Interaction between the molecule and the electric field of the radiation requires that the two ends of the molecule are not electrically identical.

Invitation to Physical Chemistry

(2) Rotational angular momentum can only change by one unit ($\Delta J = \pm 1$) during a transition. This rule conserves the combined angular momentum of radiation and the molecule.

The arrows in Fig. 2.5 indicate a few transitions allowed by the selection rules. The spectrum consists of two branches separated by a gap. The branch on the high-energy side is called the R branch; the bands in this branch correspond to $\Delta J = +1$ transitions. The other branch, called the P branch, is a collection of bands with $\Delta J = -1$.

Rotational energies are given by Eq. (2.13) and vibrational energies by Eq. (2.6). Hence R-branch bands have the energies

$$\Delta E = (h\nu + 2B), \quad (h\nu + 4B), \quad (h\nu + 6B)\ldots \tag{2.15}$$

and P-branch bands

$$\Delta E = (h\nu - 2B), \quad (h\nu - 4B), \quad (h\nu - 6B)\ldots \tag{2.16}$$

Thus the neighboring bands in each branch are separated by $2B$ and the origins of the two branches, by $4B$. The center of the two branches is where all $\Delta J = 0$ transitions would be, were they allowed. This is the missing Q branch of the spectrum. (The Q branch is prominent in some cases as we will see later.) Table 2.1 gives the wave numbers for several transitions in the hydrogen chloride rotational-vibrational spectrum.

Table 2.1. A few lines in the vibrational-rotational spectrum of the ^{35}HCl molecule.

R branch	Position in cm^{-1}	P branch	Position in cm^{-1}
$R(4)$	2980.90	$P(1)$	2865.09
$R(3)$	2963.24	$P(2)$	2843.56
$R(2)$	2944.89	$P(3)$	2821.49
$R(1)$	2925.78	$P(4)$	2798.78
$R(0)$	2906.26	$P(5)$	2775.79

Bond length of $H^{35}Cl$

After combining Eqs. (2.11) and (2.13) and rearranging we get

$$\frac{1}{r_R^2} = \frac{8\pi^2 \bar{B} m_r c}{h} \qquad (2.17)$$

$\bar{B} = 41.16[\text{cm}^{-1}]/4 = 10.29\,\text{cm}^{-1}$

$m_r = 1.627 \times 10^{-27}\,[\text{kg}]$

$c = 2.998 \times 10^{10}\,[\text{cm}\,\text{s}^{-1}]$

$h = 6.626 \times 10^{34}\,[\text{J}\,\text{s}]$

If we substitute these values into Eq. (2.17) we get

$$\left\langle \frac{1}{r_R^2} \right\rangle = 59.63 \times 10^{18}\,[\text{m}^{-2}]$$

What we have obtained from the spectrum is not the bond length but the *expectation value* of r_R^{-2}. This is because the molecule vibrates and does not have a fixed bond length. To deduce the bond length we make the approximation

$$r_e \approx \sqrt{\frac{1}{\langle 1/r_R^2 \rangle}} \qquad (2.18)$$

In this particular case we will get $1.29 \times 10^{-10}\,\text{m}$ (129 pm) as the bond length. Equation (2.18) introduces no more than half a percent error for molecules that behave like harmonic oscillators, i.e. those with nearly equidistant vibrational energy levels.

Centrifugal stretching

If HCl were a true rigid rotor, the separation between neighboring bands would be constant. As Table 2.1 indicates, however, this is not the case. As the J values increase, centrifugal stretching (in both $v = 0$ and $v = 1$) increases the bond length.

Isotopes

The splitting of rotational bands in Fig. 2.4 is due to different isotopes of chlorine. Spectroscopy has been an important tool in identifying isotopes. The classic work is by Urey who discovered ^2H (deuterium) by its faint atomic emission lines in the spectrum of atomic hydrogen. In many cases we do not have to separate the isotopic species to study their individual spectra. Their relative abundance can be estimated from the intensity of their spectral lines.

Intensities of Rotational Bands

According to the Boltzmann equation (derived in Section 9.1) the relative numbers of molecules in rotational energy levels are given by

$$\boxed{\frac{N_J}{N_{J'}} = \frac{2J+1}{2J'+1} \exp\{(E_{J'} - E_J)/k_B T\}} \qquad (2.19)$$

where k_B is the Boltzmann constant $(1.381 \times 10^{-23}\,JK^{-1})$. The intensity of a spectral band is proportional to the population in the level from which the transition arises.

To illustrate, let us estimate the relative intensities of the $R(1)$ and $R(0)$ bands (which originate from the $J=1$ and 0 levels respectively) in the $H^{35}Cl$ spectrum at 300 K. The energy difference between these two levels is $2B = 3.96 \times 10^{-22}\,J$. From Eq. (2.19) we have

$$\frac{I[R(1)]}{I[R(0)]} = \frac{3}{1} \exp\left(-\frac{3.96 \times 10^{-22}}{1.38 \times 10^{-23} \times 300}\right) = 2.73$$

The above procedure may be reversed to calculate the temperature of a sample from spectral intensities. Spectroscopy and the Boltzmann equation have been extensively used for estimating temperature of celestial bodies, plasmas, flames and high vacuum systems.

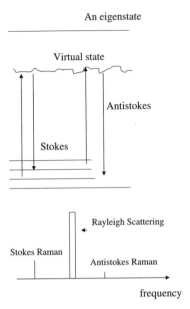

Fig. 2.6. Raman scattering is a two-photon process between a virtual state and an eigenstate. Due to the short time of photon–molecule interaction, energy levels of the molecules broaden (the uncertainty principle). A virtual state persists only during the interaction between the photon and the molecule. In Raman scattering, a photon creates a virtual state from which another photon is scattered. In the stokes bands, the molecule loses a quantum of vibrational energy; in the antistokes transitions it gains a quantum of vibrational energy. The lower part of the figure shows that the antistokes band is weaker. It arises from an upper vibrational level where the population is lower. The intense central peak is due to Rayleigh scattering.

2.3.2. *Raman spectra of homonuclear diatomic molecules: nitrogen and fluorine*

Raman scattering offers the most convenient way of investigating the spectra of homonuclear diatomic molecules and highly symmetric molecules, which are not infrared active. The principle behind Raman scattering is illustrated in Fig. 2.6.

When radiation of frequency ω_o falls on a molecule, it polarizes its charge distribution. The induced dipole moment oscillates at the impinging frequency. According to classical theory, the oscillating dipole radiates (scatters) at the frequency of oscillation, in this case at ω_o. This process is called Rayleigh scattering. The scattered radiation has the same frequency as the impinging radiation, even though it is much lower in intensity.

Rayleigh scattering is an elastic process; the molecule is a passive participant. We know, however, that molecules rotate and vibrate at particular frequencies, ω_M. Hence molecular polarizability and resultant scattering is governed by the impinging frequency and also by molecular frequencies. As a consequence the scattered light will have frequencies $\omega_o \pm \omega_M$ as illustrated in Fig. 2.6.

Raman scattering is very weak compared to Rayleigh scattering. However, with development of nearly monochromatic lasers and very low intensity detection techniques, Raman spectra have become more accessible. Unlike infrared absorption, Raman studies are done in the visible and ultraviolet region. Thus we have a much wider choice of solvents, light sources and detectors.

Figure 2.7 gives the gas phase rotational Raman spectrum of $^{19}F_2$, a molecule whose spectrum cannot be studied by infrared absorption. Raman transitions involve two photons. As a result the rotational selection rules are

$$\Delta J = \pm 2$$

This means that the rotational peaks in Fig. 2.7 are separated by $4B$. The spectrum in Fig. 2.7 shows that the peaks originating from odd J levels are more intense than the ones originating from even J levels. This is because ^{19}F has nuclear spin components of $\pm (1/2)\hbar$. We have just seen how high-resolution spectroscopy also gives information on nuclear properties. An explanation of nuclear spin effects on spectral intensities, however, will have to wait until the next chapter.

2.3.3. *Hydrogen molecules*

A hydrogen molecule has no dipole moment, and so should not absorb in the infrared region. Or, to put it in the language of spectroscopy, vibrational and rotational transitions in H_2 are forbidden transitions. They can appear as weak signals, however, if the selection rules break down. Spectroscopists have a fascination with forbidden transitions and work very hard to measure them. Herzberg in 1949 observed the vibrational-rotational spectrum of H_2 at 1 atm in a multiple reflection cell with an optical path length of 10 km. From that he was able to determine an accurate bond length for this molecule.

This heroic experiment worked thanks to the fact that, while the hydrogen molecule does not have a dipole moment, it does have a quadrupole moment. Figure 2.8 gives an idea how a quadrupole looks. The interaction between radiation with quadrupole moments is orders of magnitude weaker than the interaction with dipole moments, but as Herzberg showed, it is possible to detect.

2.4. Triatomic Molecules: Carbon Dioxide and Water

Figure 2.9 shows what are called the **normal modes** of vibration for CO_2 with the standard notation, ν_1, ν_2, ν_3. Each normal mode is a coupled displacement of atoms in a molecule. The Schrödinger

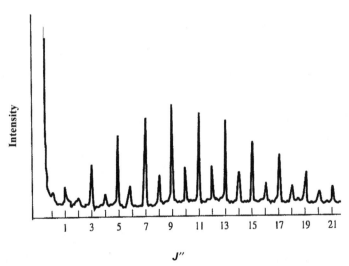

Fig. 2.7. Rotational Raman spectrum of $^{19}F_2$. The selection rule for allowed transitions is $\Delta J = \pm 2$. The numbers in the figure are the quantum numbers of the initial rotational levels. Lines arising from odd J levels are more intense than those arising from even J levels. Explanation for this is given in Section 3.5.

equation for vibrational degrees of freedom, a partial differential equation, can be separated into ordinary differential equations only when normal modes are used to represent the relative displacements of atoms. Once that is done we can think of the molecule as a collection of oscillators (harmonic in idealization) and apply the theory that we have already explored.

The vibrations ν_1 and ν_3 involve the displacement of atoms along the internuclear axis. The spectra of these vibrations are called parallel bands. CO_2 has no dipole moment and the first mode does not change that situation, hence it does not appear in the infrared spectrum. In the ν_3 mode, however, the molecule acquires a dipole moment and the mode is infrared-active. In molecules such as CO_2 with a center of inversion, infrared and Raman bands are complementary: a given mode is active either in the infrared absorption or in Raman scattering, but not in both.

The rotational selection rule for parallel bands is

$$\Delta J = \pm 1 (\text{infrared}); \quad \Delta J = 0, \ \pm 2 (\text{Raman})$$

The ν_2 mode transition is an example of a perpendicular band, where atomic displacements are perpendicular to the inter-nuclear axis. Figure 2.9 shows that this is a degenerate mode with vibrations in the perpendicular planes. Since displacements in ν_2 mode are to

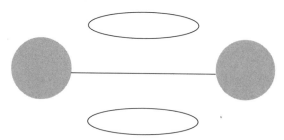

Fig. 2.8. A picture of a quadrupole in a homonuclear diatomic molecule. The shaded area represents the nuclei with positive charge. The ovals represent the electron charge clouds. The centers of positive and negative charges fall at the same point; hence the molecule has no dipole moment. However it has a quadrupole moment because the two positive charges and the two negative charges are spacially separated.

both sides of the equilibrium position, the average geometry of CO_2 is still linear.

The selection rules for perpendicular bands are: $\Delta J = 0, \pm 1$. Hence the Q branch ($\Delta J = 0$ transitions) is now allowed, as illustrated in Fig. 2.10.

Table 2.2 gives the frequencies of prominent vibrational bands in the infrared and Raman spectrum of CO_2. The band at $1285.5 \, cm^{-1}$

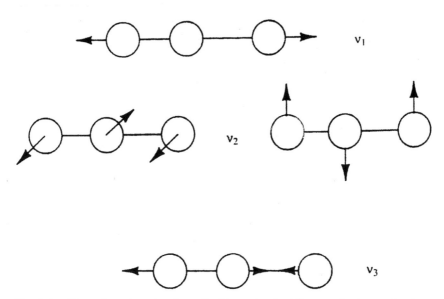

Fig. 2.9. Normal modes of carbon dioxide molecule. The standard notation is to identify the different modes with the symbol ν and a subscript. Notice that the ν_2 mode is doubly degenerate.

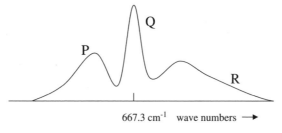

667.3 cm^{-1} wave numbers →

Fig. 2.10. A sketch of the low-resolution spectrum of the ν_2 band. The Q branch consists of several transitions with $\Delta J = 0 (0 \rightarrow 0, 1 \rightarrow 1, 2 \rightarrow 2 \ldots)$.

Table 2.2.   Prominent vibrational bands in the spectrum of carbon dioxide. The position of the second harmonic of ν_2 is anamolous because of the interaction of this state with the first harmonic of ν_1. Such interactions are very common in the vibrational spectra of polyatomic molecules.

ν_1	1388.3	Sym stretch	←O—C—O→	R very strong
ν_2	667.3	Bending	O—C—O	*ir* strong
ν_3	2349.3	Assym. stretch	O→ ←C O→	*ir* very strong
$2\nu_2$:	1285.5	Second harmonic		R very strong

R: Active only in Raman scattering; *ir*: infrared active.

contains the transitions from $v = 0$ to $v = 2$ of the ν_2 mode, and is called the second harmonic. In a molecule that is a true harmonic oscillator, this band should not appear. Anharmonicity corrections would predict a weak band at approximately $2 \times 667.3 = 1334.6 \, \text{cm}^{-1}$, but here the band is quite strong. Both this intensity and the anomalous position can be explained by the interaction of the $2\nu_2$ level with ν_1. Interactions of this sort are called Fermi resonances.

This brief foray into the CO_2 spectrum shows how the coupling between different modes plays a very crucial role. In molecules with a large number of vibrational modes, coupling could lead to what are identified as local modes: vibrations associated with chemical bonds such as C=O and C−H bonds. Thus vibrational spectroscopy is used as a fingerprint for large molecules.

Figure 2.11 shows the normal modes for a water molecule and prominent transitions in the infrared. Unlike in CO_2, the ν_2 mode in H_2O is not degenerate. In nonlinear triatomic molecules, the two degenerate vibrations interact to give a rotational mode and a vibrational mode. Hence the four vibrational degrees of freedom in a linear triatomic molecule reduce to three in a bent molecule. Figure 2.12 shows how this comes about.

2.5. Angular Momentum and the Dipole Moments

For spectroscopists, angular momentum is a more useful quantity than energy. Energy calculations are not always reliable when

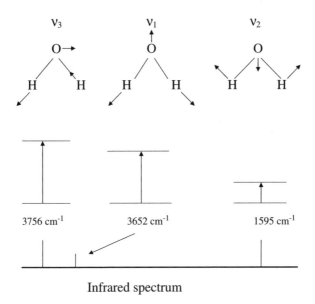

Infrared spectrum

Fig. 2.11. Normal modes of a water molecule. The arrows with atoms are not drawn to scale; displacement of the oxygen atom is much less than that of the hydrogen atoms. Below the molecules are shown $v = 0$ and $v = 1$ levels for each mode and the transitions between those levels. The spectrum at the bottom shows the relative intensities of the transitions.

there is coupling between different states as we have seen in the $2\nu_2$ band of CO_2. At higher energies, coupling becomes far more pronounced and the calculated and measured energies could differ widely. Angular momentum, however, continues to be a reliable guide in assigning transitions, since the coupling is restricted to states with the same angular momentum. In the jargon of spectroscopists, angular momentum remains a "good" quantum number.

Let us see how angular momentum provides useful information. Quantum theory restricts the orientation of the angular momentum vector to a few discrete values. This is because, besides total angular momentum, one of its three special components is also quantized. Let us call this, quite arbitrarily, the z-component. The corresponding quantum number is usually designated as M_J. Until an electric or magnetic field is applied, we will not see the effects of having different M_J values.

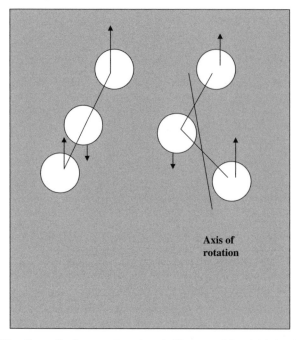

Fig. 2.12.   The three displacement vectors in linear and bent triatomic molecules.
The same displacement vectors in a bent triatomic molecule impart angular
momentum around the axis shown. In quantum mechanical treatment of this
problem, the degenerate ν_2 modes of a linear molecule are coupled to give one
vibrational mode and one rotational mode in a bent molecule.

If the total angular momentum is $\hbar\sqrt{J(J+1)}$, $J = 0, 1, 2\ldots$,
the z-component is restricted to $0\hbar$, $1\hbar$, $2\hbar\ldots \pm J\hbar$. Thus if $J = 2$,
the allowed values for M_J are

$$M_J = 2, 1, 0, -1, -2$$

Since J and M_J are related in this way, the angle between the angular
momentum vector and its z-component must be

$$\cos\vartheta = \frac{M_J}{\sqrt{J(J+1)}} \tag{2.20}$$

Thus, as shown in Fig. 2.13, the $J = 2$ vector is tilted from the
z-axis by the following angles: $23.57°$, $63.44°$, $90°$, $116.57°$, $153.44°$.
Figure 2.13 also shows the component of angular momentum, $\hbar M_J$,

z-component

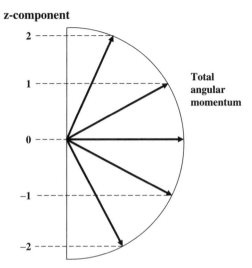

Fig. 2.13. Different possible orientations of the vector with angular momentum of $\hbar\sqrt{2(2+1)}$. The projection on the applied field axis, usually designated as the z-axis, takes the values $2\hbar$, $1\hbar$, $0\hbar$, $-1\hbar$, $-2\hbar$.

along the direction of the applied field. The J vector does not stay stationary but precesses around the z-axis and never completely lines up with it, even though it comes close at high J values (the smallest angle is $1°$ when $J = 6500$).

In the absence of imposed perturbation, each component will have the same energy since only J, and not M_J, appears in the energy formula Eq. (2.13). Indeed, there is no way of identifying any particular axis as the z-axis. This situation changes when you apply an electric (Stark effect) or magnetic (Zeeman effect) field. Figure 2.14 shows the $J = 1$ and $J = 2$ levels before and after applying the magnetic field.

Stark splitting depends on the dipole moment of the molecule. An electric field helps in the assignment of spectral lines as well as in determining molecular dipoles. For linear molecules, the J vector has to be perpendicular to the internuclear bond (see Fig. 2.15.) As a result, the projection of dipole moment on the J axis and from there onto the field axis vanishes. However, the field induces a dipole moment by distorting the charge distribution. The induced dipole

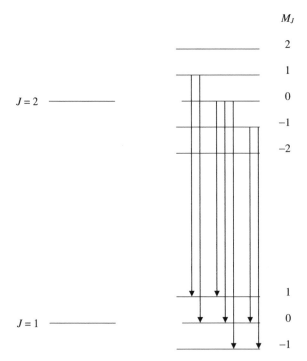

Fig. 2.14. Effect of magnetic field (Zeeman effect) on the angular momentum states. The selection rules for transitions are $\Delta M_J = 0, \pm 1$. Allowed transitions are shown in the figure. The spacing between neighboring M_J in the two J levels needs not be the same.

moment then interacts with the field. This is second order Stark splitting that is determined by the square of the electric field.

The Stark effect gives the magnitude of the dipole moment but not the direction. We depend on quantum calculations of the molecule's electronic structure (or, in favorable cases, on chemical intuition) to determine the direction of the moment.

2.6. Interstellar Molecules

Interstellar clouds, containing dust particles, appear as dark patches in the vast spaces between the stars. Among them are giant molecular

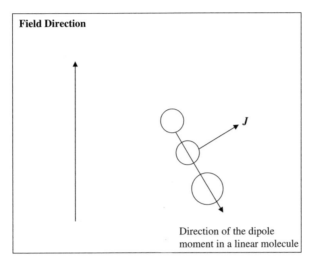

Fig. 2.15. Dipole moment of a linear molecule along the internuclear axis. Rotational angular momentum is perpendicular to this axis, hence the projection of dipole moment on the J-axis vanishes.

clouds with temperatures in the 10–20 K range and extreme low pressure ($\sim 10^{-10}$ atm). Spectroscopists have found close to 200 organic and inorganic molecules and ions in these clouds. The first, discovered between 1937 and 1940, were CH, CH^+ and CN. They were detected as absorption lines in the starlight coming through the clouds. Since the late 1960s, when the microwave and radiowave region became accessible, many more molecules and ions have been identified through rotational spectra, whose theory we have discussed above. The formation and reactions of these molecules in an environment far different from ours offer fresh challenges for astrophysicists and physical chemists.

I make this passing reference to interstellar molecules, ignoring all the important details, only to emphasize the point that our knowledge of the external world comes mainly from the interaction of radiation with matter. Spectroscopy is the most direct method of investigating that interaction.

The main themes of the chapter have been rotational and rotational-vibrational spectra of diatomic and triatomic molecules. Our discussion has been largely confined to molecules that behave like harmonic oscillators and rigid rotors, with some fruitful discussion of bond lengths and dipole moments.

Related material on the disc.

With CD-ROM

Chapter 3

Atoms

It would appear that many of the secrets of chemistry and materials science lie in the orbitals of the electron in the hydrogen atom. An **orbital** is a function of the three spatial coordinates of a *single* electron (the wave function for a harmonic oscillator is *not* an orbital unless the oscillator happens to be an electron). We use the orbitals in the hydrogen atom as they are, distort them, add them or make products of them to explain a host of chemical phenomena.

3.1. The Hydrogen Atom

The Schrödinger equation for the hydrogen atom is a partial differential equation. It can be solved after being separated into three ordinary differential equations in spherical polar coordinates (Fig. 3.1).

3.1.1. Orbital description

Table 3.1 gives four out of many solutions to the Schrödinger equation for hydrogen-like atoms, solutions we will be using in the bonding theory in Chapter 4. A familiarity with the shapes and symmetries of these functions is essential to understand chemical bonds. So, let us consider them before getting into the mathematical aspects.

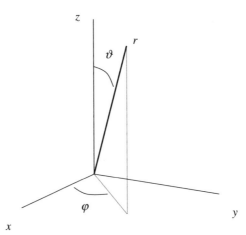

Fig. 3.1. Spherical polar coordinate system and its relation to the Cartesian axes. $z = r \cos \vartheta$; $y = r \sin \vartheta \sin \varphi$; $x = r \sin \vartheta \cos \varphi$; $dV = r^2 dr \sin \vartheta \, d\vartheta \, d\varphi$.

Table 3.1. $1s$ and $2p$ orbitals of electron in hydrogen-like atoms.

$1s$: $n = 1, l = 0, m_l = 0$	$\psi_{1s} = \dfrac{1}{\sqrt{\pi}} \left(\dfrac{Z}{a_o}\right)^{3/2} \exp\left(-\dfrac{Zr}{a_o}\right)$
$2p_z$: $n = 2, l = 1, m_l = 0$	$\psi_{2pz} = \dfrac{1}{\sqrt{32\pi}} \left(\dfrac{Z}{a_o}\right)^{3/2} \dfrac{Zr}{a_o} \exp\left(-\dfrac{Zr}{2a_o}\right) \cos \vartheta$
$2p_x$: $n = 2, l = 1, m_l = \pm 1$	$\psi_{2px} = \dfrac{1}{\sqrt{32\pi}} \left(\dfrac{Z}{a_o}\right)^{3/2} \dfrac{Zr}{a_o} \exp\left(-\dfrac{Zr}{2a_o}\right) \sin \vartheta \cos \varphi$
$2p_y$: $n = 2, l = 1, m_l = \pm 1$	$\psi_{2py} = \dfrac{1}{\sqrt{32\pi}} \left(\dfrac{Z}{a_o}\right)^{3/2} \dfrac{Zr}{a_o} \exp\left(-\dfrac{Zr}{2a_o}\right) \sin \vartheta \sin \varphi$

Complex representation of $2p$ orbitals with $m_l = 1$ and $m_l = 2$

$2p_{\pm 1}$	$\psi_{2p\pm 1} = \dfrac{1}{\sqrt{64\pi}} \left(\dfrac{Z}{a_o}\right)^{3/2} \dfrac{Zr}{a_o} \exp\left(-\dfrac{Zr}{2a_o}\right) \sin \vartheta \exp(\pm i\varphi)$

a_o: Bohr radius, 52.9 pm; $Z = 1$ for H, 2 for He$^+$, 3 for Li^{++} ...

$1s$ Function

Any picture we can draw of this or any other orbital can never show the whole picture. Since an orbital is a function of three dimensions, we would need four dimensions (three for the variables and another for the amplitude) for the correct representation. Instead, the best

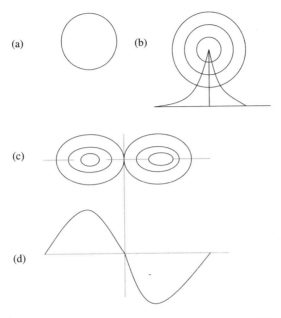

Fig. 3.2.   Graphical representations of *s* and *p*-orbitals.

we can do is to consider different representations, each indicating a significant aspect of the orbital. Figure 3.2 gives three different ways to look at this function, each useful in a particular context. Fig. 3.2(a) indicates the shape of the orbital, with the understanding that the line represents the *boundary of a sphere*. This can be made quantitative by drawing boundaries enclosing specified charge densities as shown in Fig. 3.2(b), similar to a contour map. Figure 3.2(b) gives yet another representation, a cross-section of the function in a particular plane, indicating how the orbital amplitude falls off from the center as the distance increases. If you happened to be on top of the nucleus, you would notice a steep drop in all directions. The largest amplitude for the 1*s* function is at the center of the nucleus. The nucleus is not solid; it too is a wave packet.

$2p_z$ Function

We have to consider another factor now: the phase, indicated by a positive or negative sign. Phase is not a permanent fixture.

In time-dependent theory, phase oscillates with a frequency $\left(\frac{Et}{\hbar}\right)$. Figure 3.2(c) represents the shape of the orbital as three contour lines. Figure 3.2(d) is the amplitude of the function along the z-axis. If you were on the nucleus and looking in the z-direction, you would see a hill on one side and a valley on the other. You would see nothing along either of the perpendicular directions. This figure is also the cross-section of the $2p_z$ orbital in the x–y plane.

Symmetry

We also need to be familiar with the symmetry properties of the wave functions in order for bonding theory to make sense.

We all have an instinctive feeling about symmetry. Quantifying it by defining the **symmetry operators** will help us understand molecular structure. All of the symmetry operators that we will consider leave the wave function indistinguishable from the original configuration except for the phase. This is a very general definition, but one that serves our purpose. The effects of the following operators on the hydrogen orbitals are sketched in Fig. 3.3:

$\hat{i}$ Inverts an orbital through the center

$\hat{C}_{2z}$ Rotates an orbital around the z-axis by $180°$

$\hat{C}_{2y}$ Rotates an orbital around the y-axis by $180°$

$\hat{\sigma}_{xy}$ Reflects an orbital in the xy plane drawn at the origin

$\hat{\sigma}_{xz}$ Reflects an orbital in the xz plane drawn at the origin

Consider first the $1s$ orbital. Since it is spherically symmetric, the above operations leave it unchanged. So, following the quantum theoreticians' penchant for eigenvalue equations, we write

$$\hat{O}(1s) = 1(1s) \tag{3.1}$$

where $\hat{O}$ is one of the symmetry operators listed above. For $2p_z$ the eigenvalues are $+1$ for $\hat{C}_{2z}$, $\hat{\sigma}_{xz}$ and $\hat{\sigma}_{yz}$ operators and -1 for the others. You should verify this and familiarize yourself with symmetry operators.

It is not purely accidental that the hydrogen orbitals conform to Eq. (3.1). This is a consequence of symmetry operators **commuting**

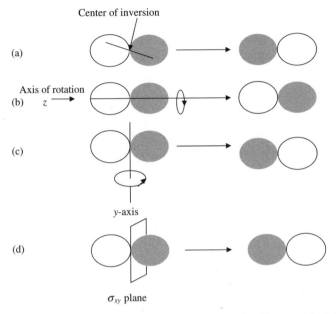

Fig. 3.3. Different symmetry operations on p-orbitals. The s-orbital is totally symmetric for all the operations listed: (a) inversion, (b) rotation around z-axis, (c) rotation around y-axis and (d) reflection in the x–y plane.

with the Hamiltonian, the operator for energy. There is an underlying principle here which states that, if two operators commute, eigenfunctions of one operator must also be the eigenfunctions of the other operator. Operators are only said to commute if the order in which they are applied does not matter. Thus operators $\hat{A}$ and $\hat{B}$ commute if

$$\hat{A}\hat{B}\psi = \hat{B}\hat{A}\psi \qquad (3.2)$$

This equation is also written as

$$\{\hat{A}\hat{B} - \hat{B}\hat{A}\} = 0 \qquad (3.3)$$

where the eigenfunction is supposed to be hiding behind the curtain or simply as

$$\{\hat{A}, \hat{B}\} = 0 \qquad (3.4)$$

To complete the argument, we must show that the symmetry operators and the Hamiltonian commute. That turns out to be quite simple. Consider the following fragment of an equation,

$$\hat{i}(\hat{H}\psi) =?$$ (3.5)

The only thing that the operator $\hat{i}$ does is to change the point at x, y, z to a point at $-x$, $-y$, $-z$ in the Hamiltonian and also in the orbital. If the electron were to be at x, y, z, its position would change to an equidistant position at $-x$, $-y$, $-z$. The energy, and hence the Hamiltonian, are not altered. Formally stated:

$$\hat{i}(\hat{H}\psi) = \hat{H}(\hat{i}\psi) = E(\hat{i}\psi)$$ (3.6)

The same holds for the rest of the symmetry operators.

3.1.2. Angular momentum

Symmetry and angular momentum are more useful than energy in assigning spectroscopic transitions. Think of angular momentum operators as analogs of symmetry operators.

Let us go ahead and familiarize ourselves with the properties of angular momentum operators. The Hamiltonian operator in the Schrödinger equation for hydrogen, $\hat{H}$, is the sum of kinetic energy and potential energy operators. It commutes with $\hat{l}^2$, the operator for the square of the total angular momentum. $\hat{H}$ and $\hat{l}^2$ also commute with the operators for each of the three components of angular momentum, $\hat{l}_x$, $\hat{l}_y$ and $\hat{l}_z$. The proof for commutation, of course, should be more substantial than the comments here.

However, $\hat{l}_x$, $\hat{l}_y$ and $\hat{l}_z$ do not commute among themselves. Hence the hydrogen orbitals are eigenfunctions of $\hat{l}^2$ and just one of the components, usually identified as $\hat{l}_z$, but not of the other two components.

In the abbreviated notation of Eq. (3.4)

$$\{\hat{l}^2, \hat{H}\} = 0; \quad \{\hat{l}_z, \hat{H}\} = 0; \quad \{\hat{l}^2, \hat{l}_z\} = 0$$ (3.7)

However,

$$\{\hat{l}_x, \hat{l}_y\} \neq 0; \quad \{\hat{l}_y, \hat{l}_z\} \neq 0; \quad \{\hat{l}_z, \hat{l}_x\} \neq 0$$ (3.8)

Table 3.1 lists the angular momentum eigenvalues for each of the hydrogen orbitals. In the standard format, the eigenvalue equations are

$$\hat{l}^2\psi = \hbar^2 l(l+1)\psi \tag{3.9}$$

and

$$\hat{l}_z\psi = \hbar m_l\psi \tag{3.10}$$

Section 2.5 contains similar equations for molecular rotations. The same symbol, l, is used for operators and eigenvalues in the first equation. Operators are distinguished by circumflexes. Keep in mind that if you read "angular momentum is 2", either here or in other books, it means that the angular momentum is actually $\hbar\sqrt{2(2+1)}$ and its z-component has one of the following values: $2\hbar, 1\hbar, 0\hbar, -1\hbar, -2\hbar$. Figure 2.13 gives the orientations of the angular momentum vector.

Note that in Section 2.3 we were able to obtain the energies of the rigid rotor from the angular momentum values because of the commutation relation.

3.1.3. *Energy*

Orbital energies depend on the **principle quantum number**, n, as follows:

$$E_n = \frac{1}{4\pi\varepsilon_o}\left(\frac{Ze^2}{2a_o n^2}\right) \tag{3.11}$$

where ε_o is the permittivity of vacuum $(C^2\,J^{-1}m^{-1})$, e is the elementary charge and Ze is the charge on the nucleus. The bracketed term is one-half the potential energy of the electron at a distance of $a_o n^2$. The **Bohr radius**, $a_o = 52.9\,\text{pm}$, is given by the equation

$$a_o = \frac{\hbar^2\varepsilon_o}{\pi m_r e^2} \tag{3.12}$$

Figure 3.4 shows the electron energies and some of the various transitions observed in the spectrum.

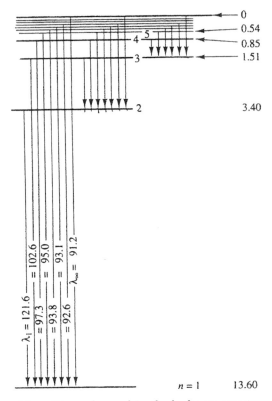

Fig. 3.4. Several transitions observed in the hydrogen spectrum. There are no selection rules for the n quantum number. The numbers on the right are the energies in electron-volts expressed as positive quantities.

Much of our discussion on bonding depends solely on the symmetry of the orbitals. Nevertheless some familiarity with mathematical aspects of the theory, as illustrated in the following examples, will be helpful.

Orthogonality

Functions f_i and f_j are said to be orthogonal if $\int f_i^* f_j dV = 0$. In this equation f_i^* is the complex-conjugate of f_i and dV is the volume element, $dV = r^2 dr \sin \vartheta \, d\vartheta \, d\varphi$. We don't always work with functions having an imaginary part, in which case a function and its complex conjugate would be identical (see Eq. (1.29) for an example).

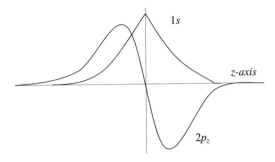

Fig. 3.5. The integral $\int 1s2p_z dV$ is the area under the $1s2p_z$ product function. On the left side of the vertical line, the two functions overlap to give a positive value for the area. On the right side, the two functions overlap to give a negative value for the area. Since the sum of the two areas is zero, the overlap integral is zero.

To illustrate, consider the $1s$ and $2p_z$ orbitals. The condition for orthogonality is

$$\int 1s2p_z dV = 0$$

We can integrate this expression to show that the integral goes to zero, but it is simpler to use a graphical method. Figure 3.5 sketches $1s$ and $2p_z$ functions along the z-axis. Note that the integral is a product of an even $[f(z) = f(-z)]$ and an odd $[f(z) = -f(-z)]$ function and so has to vanish.

A rough sketch on a piece of paper is all that is needed to simplify some quantum calculations. Nevertheless, let us evaluate the above integral to gain familiarity with the hydrogen orbitals. With the functions given in Table 3.1, we expand the above integral:

$$\int 1s2p_z dV = \underbrace{\frac{1}{\sqrt{\pi}} \left(\frac{Z}{a_o}\right)^{3/2} \frac{1}{4\sqrt{2\pi}} \left(\frac{Z}{a_o}\right)^{3/2}}_{\text{normalization constants}}$$

$$\times \iiint \underbrace{\left(\exp\left(-\frac{Zr}{a_o}\right)\right)}_{1s} \underbrace{\left(\left(\frac{Zr}{a_o}\right) \exp\left(-\frac{Zr}{2a_o}\right) \cos\vartheta\right)}_{2p_z} \underbrace{r^2 dr \sin\vartheta d\vartheta d\varphi}_{dV}$$

and separate into three integrals,

$$Integral = \frac{1}{\sqrt{\pi}} \left(\frac{Z}{a_o}\right)^{3/2} \frac{1}{4\sqrt{2\pi}} \left(\frac{Z}{a_o}\right)^{3/2} \int \left(\exp\left(-\frac{Zr}{a_o}\right)\right) \left(\frac{Zr}{a_o}\right)$$

$$\times \exp\left(-\frac{Zr}{2a_o}\right) r^2 dr \int \cos\vartheta \sin\vartheta \, d\vartheta \int d\varphi$$

The limits on integration are 0 and π for the ϑ integral. This integral goes to zero. Hence we need not evaluate the other two integrals to show that $1s$ and $2p_z$ functions are orthogonal. While this exercise gives us some practice with the mathematical aspects of hydrogen functions, for the most part we get the same results with symmetry as a guide, as Fig. 3.5 illustrates.

Most Probable Distance of $1s$ Electron

The most probable distance occurs where there is maximum probability density. Let us consider the $1s$ function for illustration. We want the probability only as a function of radial distance. If we square the wave function and integrate over angle variables, we obtain the radial probability density.

$$\rho(r)dr = \frac{1}{\pi} \left(\frac{Z}{a_o}\right)^3 \int_\theta \int_\varphi \exp\left(-\frac{2Zr}{a_o}\right) r^2 dr \sin\theta \, d\theta \, d\varphi$$

$$= 4\left(\frac{Z}{a_o}\right)^3 \exp\left(-\frac{2Zr}{a_o}\right) r^2 dr$$

at the maximum,

$$\left(\frac{d\rho(r)}{dr}\right) = 0$$

from this we get:

$$r(\text{most probable}) = \frac{a_o}{Z}$$

Average Distance of 1s Electron

The operator for distance from the nucleus is $\hat{r}$. Hence the average distance of the electron is obtained from the formula for expectation value.

$$\langle r \rangle = \frac{1}{\pi} \left(\frac{Z}{a_o} \right)^3 \int_0^{2\pi} d\varphi \int_0^\pi \sin \theta \, d\theta \int_0^\infty \exp \left(-\frac{Zr}{a_o} \right) \hat{r} \exp \left(-\frac{Zr}{a_o} \right) r^2 dr$$

$$= \frac{3a_o}{2Z}$$

3.2. Spin

Even before Schrödinger developed the theory of the hydrogen atom, it was clear that another quantum number was needed to explain its spectrum. Figure 3.6 shows energy levels in the $4p \rightarrow 3s$ spectrum of hydrogen. The Schrödinger theory predicts just one line for this

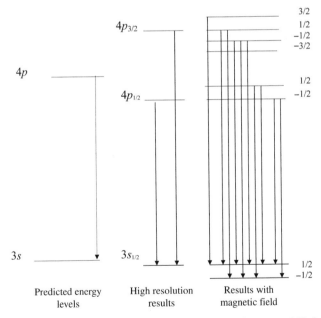

Fig. 3.6. The energy levels are predicted by the Schrödinger and Bohr theories. Both theories predict only one transition (left). However, at high resolution the spectrum shows two lines which can only be explained by invoking spin angular momentum for the electron (middle). When a magnetic field is applied (Zeeman effect) several transitions are observed (right).

transition but what is actually observed is a **doublet**, a pair of closely spaced transitions.

Incontrovertible experimental results show that the electron has a **magnetic moment** (just like a compass needle) and its z-component can orient along one of **two opposite directions** in space. Built on these results comes a theory that fits the rest of quantum mechanics through definitions of angular momentum, eigenvalues and commutation.

Classically, a magnetic moment is associated with spinning or rotating charged particles. So initially it was suggested that the electron spins around its axis and the two orientations of the magnetic moment correspond to the two orientations of the angular momentum vector. However, the idea of a localized particle and the surface velocity needed to account for observed magnetic moment could not be justified. Thus, although the accepted view is that both observed magnetic moment and its relation to quantized angular momentum fit into the theory nicely and explain many properties, there is no *mechanical model* for what we casually call spin. To avoid fixation with a mechanical model, spin is frequently referred to as **intrinsic angular momentum**, a term which could mean either "Don't ask any questions" or something so profound that we can never understand it.

The formal theory of spin defines operators for the square of spin angular momentum $(\hat{s}^2)$ and its z-component $(\hat{s}_z)$, and their eigenfunctions (α and β) as indicated by the following equations:

$$\hat{s}^2\alpha = \hbar^2 \left(\frac{1}{2} \left(\frac{1}{2} + 1 \right) \right) \alpha; \quad \hat{s}^2\beta = \hbar^2 \left(\frac{1}{2} \left(\frac{1}{2} + 1 \right) \right) \beta \quad (3.13)$$

$$\hat{s}_z\alpha = \left(\frac{1}{2} \right) \hbar\alpha; \quad \hat{s}_z\beta = - \left(\frac{1}{2} \right) \hbar\beta \quad (3.14)$$

Figure 3.7 shows the two possible orientations of spin angular momentum.

The interpretation of doublet structure follows from this theoretical construct. The orbital and the spin angular momenta interact

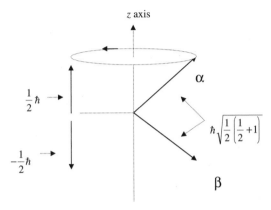

Fig. 3.7. Two possible orientations of the spin vector. The total spin angular momentum is shown on the right-hand side and the values for components on the left.

(spin−orbit coupling) and the resultant angular momentum, j, determines the structure of the spectrum.

Since angular momenta are vectors, components of two angular momenta are added to find out the resultant angular momentum. Table 3.2 shows all the components that result from adding $l = 1$ and $s = 1/2$ angular momenta. The highest value $j = 3/2$ must correspond to an angular momentum of $\hbar\sqrt{3/2(3/2+1)}$ which must

Table 3.2. As angular momenta are vectors, we must add the components to get the total angular momenta. In the following table the first row is the possible angular momentum components for a p-electron. The first column gives the components of spin.

m_l	1	0	−1	
m_s				
1/2	3/2	1/2	−1/2	Four components with $j = \frac{3}{2}$
−1/2	1/2	−1/2	−3/2	

Two components with $j = \frac{1}{2}$

have four z-components; $3/2\hbar$, $1/2\hbar$, $-1/2\hbar$, $-3/2\hbar$. The remaining two must then belong to $j = 1/2$.

The two transitions in Fig. 3.6 correspond to the transitions from two different j levels of the $4p$ function. Figure 3.6 also shows the effect of a magnetic field (Zeeman effect) on the transitions. The transitions are further split because of the different orientations of the j vector.

3.3. The Helium Atom

Helium was first discovered in the spectrum of the sun and hence the name from *helios*, the Greek word for sun. Our interest in helium stems from the fact that the atom has two electrons and thus forms a model for multi-electron systems, which mark a different direction for quantum theory. Repulsion between electrons does not allow the Schrödinger equation to be separated into ordinary differential equations. So how do we get the wave functions and energies for helium?

3.3.1. *Orbital approximation*

The most commonly used technique for multi-electron systems is the orbital approximation. Here we assume that the *wave function* for a multi-electron system is *approximated* by a product of *orbitals*. Thus

$$\Psi(1, 2, 3 \ldots) = \psi(1)\psi'(2)\psi''(3) \ldots$$

The numbers in parentheses denote the electrons. The word "approximation" simply means that the wave function is not a solution to the Schrödinger equation. The orbital approximation and its several stages of improvements have given us very valuable qualitative insights into molecular properties, as well as quantitative results that rival any exact theory.

The product of orbitals is called a **configuration**. The orbitals need not be hydrogen atom functions, although they are the most frequently used with some modification or the other. However, the configuration must satisfy *Pauli's principle*, the basis of which happens to be another symmetry requirement.

Let $\hat{P}_{12}$ be an operator that permutes electrons 1 and 2. Then

$$\hat{P}_{12}[\hat{P}_{12}\Psi(1,2,3\ldots)] = 1\Psi \qquad (3.15a)$$

In other words, the second permutation reverses the operations of the first and returns the wave function to one indistinguishable from the original. Is this a useless exercise proving an obvious point? Before rushing to a conclusion, consider the implications of Eq. (3.14). For this equation to be true, the wave function must meet the following condition:

$$\hat{P}_{12}\Psi(1,2,3\ldots) = \pm 1\Psi \qquad (3.15b)$$

A multi-particle wave function has to be either symmetric (eigenvalue: +1) or antisymmetric (eigenvalue: −1). This is the essence of the **Pauli exclusion principle**, which has to be accepted as an additional postulate in quantum theory.

There is a second part to the principle which works, but it is not easy to see why. Elementary particles with half-integer spins, e.g. electrons and ^{1}H nuclei, can only have **antisymmetric wave functions**, called Fermions. Elementary particles with integer spins, e.g. ^{2}H nuclei, can only have **symmetric wave functions**, called Bosons. The more common statement of the principle is that two electrons may occupy the same orbital only if they have opposite spin. To see the connection between the two statements let us go ahead and put two electrons with the *same spin* in 1s orbital. In that case, the configuration is $1s(1)\alpha(1)\ 1s(2)\alpha(2)$ which may be represented diagrammatically as ↑↑. The numbers in parentheses denote electrons. If you exchange the electrons, you get the same function back. That means the eigenvalue for the exchange operator is +1 and the configuration is symmetric, which violates the Pauli principle for electrons.

3.3.2. *Configurations for helium*

Any configuration must have the symmetry of the wave function it approximates. Since there are two electrons in helium, any acceptable configuration must be antisymmetric to their exchange. Thus the lowest energy configuration for helium that includes spin has to be

$$\psi = \sqrt{\frac{1}{2}}\,(1s\alpha\,(1)\,1s\beta(2) - 1s\alpha(2)1s\beta(1)) \qquad (3.16)$$

In this equation $1s\alpha$ and $1s\beta$ are the two different products of spin and orbital functions. Writing the above function after exchanging electrons results in a function that is the negative of the above. Hence Eq. (3.16) is an antisymmetric wave function. A positive sign in Eq. (3.16) makes it an unacceptable symmetric function.

It is worth noting that in constructing configurations, the orbitals need not be those of the hydrogen atom. All that is required is that they have the appropriate spherical symmetry.

Equation (3.16) is also expressed as

$$\psi = \sqrt{\frac{1}{2}}(1s(1)1s(2))(\alpha(1)\beta(2) - \alpha(2)\beta(1)) \qquad (3.17)$$

We see from this that the space part is symmetric, the spin part is antisymmetric and the product of the two is antisymmetric. Thus we were able to put two electrons into one orbital without violating Pauli's principle.

Helium is *diamagnetic* in its lowest energy state. This is a consequence of the antisymmetric nature of the above configuration which leads to zero net spin. An excited configuration for helium has one electron in the $1s$ orbital and the other electron in the $2s$ orbital. The combination of orbital and spin parts, consistent with Pauli's principle, leads to the following **terms**.

Singlet term:

$$^{1}\Phi = \sqrt{\frac{1}{2}}(1s_a(1)1s_b(2) + 1s_a(2)1s_b(1))(\alpha(1)\beta(2) - \alpha(2)\beta(1)) \qquad (3.18)$$

Triplet term:

$$^{3}\Phi = \sqrt{\frac{1}{2}}(1s_a(1)1s_b(2) - 1s_a(2)1s_b(1)) \begin{bmatrix} \alpha(1)\alpha(2) \\ \alpha(1)\beta(2) + \alpha(2)\beta(1) \\ \beta(1)\beta(2) \end{bmatrix} \qquad (3.19)$$

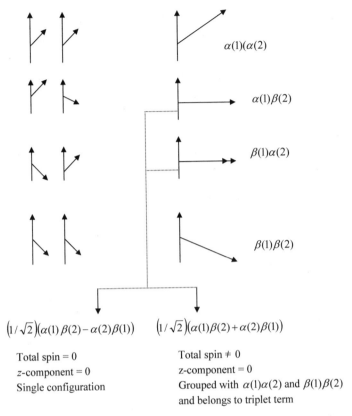

$$\left(1/\sqrt{2}\right)\!\left(\alpha(1)\,\beta(2)-\alpha(2)\beta(1)\right) \qquad \left(1/\sqrt{2}\right)\!\left(\alpha(1)\beta(2)+\alpha(2)\beta(1)\right)$$

Total spin = 0
z-component = 0
Single configuration

Total spin ≠ 0
z-component = 0
Grouped with $\alpha(1)\alpha(2)$ and $\beta(1)\beta(2)$
and belongs to triplet term

Fig. 3.8. On the left-hand side of the diagram, four possible orientations of two spins are shown. On the right-hand side, the combined spin angular momenta for two electrons are shown. The two middle functions, $\alpha(1)\beta(2)$ and $\alpha(2)\beta(1)$, are not eigenfunctions of the exchange operator unless they are mixed $1/\sqrt{2}(\alpha(1)\beta(2) + \alpha(2)\beta(1))$ is a symmetric function. Its z-component of spin is 0, but its total spin is $\hbar\sqrt{1/2(1/2 + 1)}$. Hence it belongs with the other two symmetric functions and forms part of a triplet term. $1/\sqrt{2}(\alpha(1)\beta(2) - \alpha(2)\beta(1))$ is an antisymmetric function. It belongs to the singlet term because not only its component along the z-direction but also its total spin vanishes.

For the singlet term, the symmetric orbital part is multiplied by the antisymmetric spin part. The triplet term has three states. The equations are obtained by multiplying the antisymmetric orbital part by the three symmetric spin functions. Figure 3.8 gives an explanation of how these combinations arise.

3.3.3. *Helium configuration energies*

Let us focus on the lowest energy configuration first. The Hamiltonian for the helium atom consists of (1) kinetic energy of electrons, (2) potential energy of electron–nuclear attraction and (3) potential energy of mutual repulsion of electrons. Without the third term the problem becomes trivial; the Hamiltonian can be easily separated into two hydrogen atom-like Hamiltonians and its energy is a sum of the two hydrogen-like atoms each with $Z = 2$. The configuration will simply be a product of two hydrogen orbitals, $1s(1)1s(2)$, with $Z = 2$. Unfortunately the third term, electron repulsion, cannot be ignored. How do we incorporate electron repulsion into such a simple configuration?

The concept of *screening* allows us to take into account some of the repulsion. At any given time, one electron will be closer to the nucleus than the other. So it is reasonable to assume the first electron "sees" the nucleus with $Z = 2$ while the second sees a screened nucleus with $Z < 2$. These roles, of course, are continuously reversed. The procedure then is to calculate energy using the $1s$ functions given in Table 3.1 but keeping Z as a variable

$$E(Z) = \iint 1s^2 \hat{H} 1s^2 dV_1 dV_2 \qquad (3.20)$$

and find the Z that gives minimum energy. The calculations show that $Z = 1.6875$ gives an energy just 2 per cent above the experimental value.

What has been done is a "mathematical experiment." Why should we expect this procedure to give us anything trustworthy? The answer lies in the **variation principle**. According to this principle, the trial wave function we choose cannot give an energy lower than the true energy:

$$E_{\text{trial}} > E_{\text{true}} \qquad (3.21)$$

The variation principle applies only to the lowest energy state of a given symmetry.

The variation principle helps us, because unlike with helium, we do not usually know the true energy. For helium it was obtained from the ionization of both electrons, but this is difficult with larger atoms and impossible with molecules. If we try different wave functions, we can be certain that we will never go below the true energy. The wave function that gives the lowest energy is the best we can get. We might hit the right energy, if we happen to stumble on the correct wave function.

Ultimately, quantum chemists' interest is not in energy. Who cares about the true energy of any molecule? Theoreticians use the wave functions to calculate the properties of materials. They assume that the wave function that gives the lowest energy is the best available wave function for the calculation of all properties.

Another way to improve energy (i.e. get a better wave function) is to mix (superpose) excited configurations with the lowest energy configurations. In one of these "mathematical experiments," 35 higher energy configurations were superposed on the lowest energy configuration of helium. The calculated energy was less than 0.02 per cent higher than the true energy.

Why should the calculated energy improve when we superpose more functions, as in the case of helium? How do we choose the weight of each function in a superposition? For the answers to these questions we will have to wait until Section 4.2.

3.4. Excited Terms of Helium

A precise calculation of the excited energies of an atom is out of the question since the variation principle does not provide any guidance. Instead, that guidance comes from the angular momentum operators that commute with the Hamiltonian. It is almost trivial to find the angular momentum eigenvalues.

There is a simple rule for coupling angular momenta. If l_i and l_j are two angular momentum quantum numbers and $(l_i > l_j)$, the combined angular momentum quantum number, L, will have the values: $L = l_i + l_j, l_i + l_j - 1, l_i + l_j - 2, \ldots, l_i - l_j$. For instance, if $l_i = 2$ and $l_j = 1$ are the quantum numbers, the combined angular momentum will have the quantum numbers 3, 2, 1. The number of components previous to coupling are 5 (for l_i) $\times$ 3 (for l_j) = 15; the number of components after coupling are 7 for $L = 3$, 5 for $L = 2$ and 3 for $L = 1$ or a total of 15. The numbers of components remain the same (Table 3.3).

Terms corresponding to different angular momenta can be identified unambiguously in the spectrum. The following is a brief outline of the procedure:

1. The configuration energy gives a rough idea where to expect the transitions.

2. The total orbital angular momentum of electrons is obtained by adding the angular momentum of the electrons; two in the case of helium. Thus, if the configuration is $2p3p$ ($l = 1$ for each electron), total angular momentum, L, takes the values 2, 1, 0. Terms corresponding to these L values are designated by the symbols $F(L = 3)$, $D(L = 2)$ and $P(L = 1)$.

3. Since electrons in the $2p3p$ configuration can have either parallel or antiparallel spins, we will get the following levels: ^3D, ^1D, ^3P and ^1P, ^3S and ^1S. From the multiplet structure and the selection rules for the transitions, it is possible to identify each line in the spectrum with a specific transition.

There is an important lesson to be learned here. Formal quantum theory sometimes tends to be abstract and some of its applications mathematically sophisticated, but that does not tell the full story.

Table 3.3.   Coupling of two orbital angular momenta.

m_l	2	1	0	−1	−2	← Electron 1
m_l						
1	3	2	1	0	−1	
0	2	1	0	−1	−2	
−1	1	0	−1	−2	−3	
		$j = 1$		$j = 2$	$j = 3$	

↑
Electron 2

Quantum theory has also been able to explain a large number of phenomena with only minimal mathematics, as the above example shows, and it is these qualitative aspects of quantum theory that make it so powerful. You will see in the next chapter that we can give an accurate account of bonding and some molecular properties without getting mired in mathematics.

3.5. Symmetry and Rotational Spectra

Let us revisit the $^{19}F_2$ Raman spectrum shown in Section 2.3.2. The intensity alteration observed in the spectrum is a result of Pauli's principle, which states that "particles" with half-integer spins can only have antisymmetric wave functions. The ^{19}F nucleus has half-integer spin, hence its total wave function (product of nuclear, electronic, vibrational and rotational wave functions) has to be antisymmetric.

For the fluorine molecule, the electronic wave function is symmetric because, as we will see in the next chapter, spins are paired. Its vibrational wave function when $v = 0$ is also symmetric, as you can see from the wave function shown in Fig. 2.3. That leaves us with the rotational wave function. The symmetry of the rotational wave functions is exactly the same as that of hydrogen orbitals. The hydrogen orbitals have an extra variable, the electron–nucleus separation, but it only makes orbitals trail off at a large distance and does not alter their symmetry properties. For symmetrical and angular momentum properties, we can consider the quantum numbers J and l synonymous.

By examining the sketches of the hydrogen orbitals you can readily see that even l and J values have symmetric functions while odd l and J values have antisymmetric functions.

The spin functions for two nuclei (with half-integer spins) are identical to those shown in Eqs. (3.18) and (3.19). They are either symmetric (triplets) or antisymmetric (singlet.) Since the total wave function has to be antisymmetric, we can only combine odd J functions with symmetric spin functions and even J functions with antisymmetric spin function. Thus we have

J	Components of J	Singlet(1) or triplet(3)	Total degeneracy
0	1	1	1
1	3	3	9
2	5	1	5
3	7	3	21

The antisymmetry requirement makes the odd J values more populated and transitions from them more intense.

In this chapter we have explored symmetry, angular momentum and the energies of orbitals in hydrogen-like atoms. We have also considered spin, Pauli's exclusion principle and the antisymmetric nature of configurations for atoms with many electrons. These topics form the basis for understanding bonding in molecules. The brief survey of atomic spectra shows the importance of angular momentum in developing a coherent theory. That symmetry of wave functions controls spectral intensities is dramatically illustrated by the rotational Raman spectrum of $^{19}F_2$ (Fig. 2.7).

 Related material on the disc.

With CD-ROM

Chapter 4

Molecules

Arguably, the most important symbols in chemistry are the lines between atoms denoting chemical bonds. The idea that the properties of molecules are related to their constituent atoms and the bonds between them is as old as the atomic theory. There are millions of molecules, but only a very small number of distinct bonds. Due to this few-to-many connection, we gain an insight into the properties of a vast number of compounds by understanding just a few types of bonds. For instance, there are hundreds of compounds comprised of just carbon and hydrogen atoms; we can gain at least a qualitative understanding of their properties from knowing the characteristics of only four bonds (C–H, C–C, C=C, C≡C.)

Although ideas about chemical bonds have played a significant role throughout the development of chemistry, qualitative and quantitative theories had to wait until the advent of Schrödinger's equation in 1926.

4.1. The H_2^+ Molecule

The H_2^+ molecular ion, observed in electrical discharges, provides a good starting point for a discussion of bonding. The dissociation energy of H_2^+ is 2.79 eV and its equilibrium bond length is 106 pm.

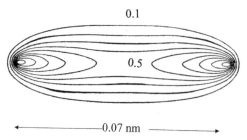

0.1

0.5

←——————————0.07 nm ——————————→

Fig. 4.1. Probability density function (Burrau, 1927) for an electron in the H_2^+ molecular ion. The two numbers next to contour lines indicate relative probability density, which falls off in steps of 0.1 from the nuclei.

Within a year of the publication of Schrödinger's theory, Burrau was able to successfully apply it to the H_2^+ molecular ion. His calculations were further refined by other theoreticians. We can learn about the nature of the chemical bond from the results of these calculations without going through the difficult technical details. Figure 4.1 gives a sketch of the lowest energy wave function obtained by numerical integration of the Schrödinger equation in prolate elliptical coordinates. If you rotate the figure around the internuclear axis, each line in the figure becomes a contour surface. What we have here is a stationary state. The orbital and the electron–charge distribution (probability density) obtained by squaring the orbital extends over the whole molecular ion. The calculated dissociation energy and bond distance are in excellent agreement with the experimental results.

4.1.1. *Molecular orbitals for H_2^+*

The wave function for H_2^+ in Fig. 4.1 is appropriately called a **molecular orbital (MO)**. An orbital, by definition, is a function of one, and only one, electron whether it is confined around an atom or spread over the whole molecule. The MO theory starts with the assumption that the wave function for *each electron* in a molecule (molecular orbital) is delocalized over the whole molecule. This idea was developed by Hund and Mulliken, who were both interested in the spectroscopy of diatomic molecules. From the spectra they were able to deduce the symmetries of molecular orbitals for the lowest energy as well as for higher energy states of many diatomic molecules.

How do we get MOs for all these states? The calculations of the type done by Burrau, heroic as they were, cannot be extended to other states of H_2^+ or other molecules. Lennard–Jones suggested a procedure that is both simple and very general. He pointed out that we can *mimic* the MOs by **linear combination of atomic orbitals (LCAO)**. For instance, we can approximate the Burrau function by linearly combining the $1s$ functions

$$\psi(\text{Burrau}) \approx A(1s_a + 1s_b) \equiv \varphi_1 \qquad (4.1)$$

Here A is the normalization constant, while $1s_a$ and $1s_b$ are the familiar hydrogen atom functions, one centered on nucleus a, and the other, on nucleus b.

A word of caution: Equation (4.1) does not imply that the electron is scuttling between the two nuclei; to say that would be absolutely wrong. We have added two familiar atomic orbitals to come up with one MO that resembles and approximates the solution to the Schrödinger equation. The two atomic orbitals form the **basis set** for the "synthesis" of one molecular orbital.

The fact that the equality in Eq. (4.1) is approximate might make you uncomfortable. How good is this approximation? We can test this by calculating the expectation value of energy.

$$\langle E \rangle = \int A^2 (1s_a + 1s_b) \hat{H} (1s_a + 1s_b) dV \qquad (4.2)$$

The results of evaluating the integrals are: dissociation energy = 1.76 eV; bond length = 132 pm. True, they differ from the experimental values (Table 4.1) by a wide margin, but we were able to get the numbers within the ballpark from a simple theory which was generated after centuries of speculation about chemical bonds.

4.1.2. *Extension of the basis set*

Of course, we need not stop here. There are ways of improving the accuracy of these calculations. Figure 4.1 shows that the charge distribution around the two nuclei is not spherically symmetric but polarized along the internuclear axis. This gives the idea that we

Table 4.1. Electronic configurations and bonding in homonuclear diatomic molecules. See text for discussion.

Molecule	Electronic configuration	Net bonding electrons	D_e/eV	r_e/pm
H_2^+	$1s\sigma_g$	1	2.79	106
H_2	$1s\sigma_g^2$	2	4.75	74
He_2^+	$1s\sigma_g^2 1s\sigma_u$	1	2.5	108
He_2	$1s\sigma_g^2 1s\sigma_u^2 = [He_2]$	0	?	?
Li_2	$[He_2]2s\sigma_g^2$	2	1.14	267
Be_2	$[He_2]2s\sigma_g^2 2s\sigma_u^2 = [Be_2]$	0	<0.1	245
B_2	$[Be_2]2p\sigma_g^2$	2	3.01	159
C_2	$[Be_2]2p\sigma_g^2 2p\pi_{ux} 2p\pi_{uy}$	4	6.36	124
N_2^+	$[Be_2]2p\sigma_g^2 2p\pi_{ux}^2 2p\pi_{uy}$	5	8.86	112
N_2	$[Be_2]2p\sigma_g^2 2p\pi_{ux}^2 2p\pi_{uy}^2 = [N_2]$	6	9.90	109
O_2^+	$[Be_2] \ 2p\sigma_g^2 2p\pi_{ux}^2 2p\pi_{uy}$	5	6.77	112
O_2	$[Be_2] \ 2p\sigma_g^2 2p\pi_{ux}^2 2p\pi_{uy}^2 2p\pi_{gx} \ 2p\pi_{gy}$	4	5.21	120
O_2^-	$[Be_2] \ 2p\sigma_g^2 2p\pi_{ux}^2 2p\pi_{uy}^2 2p\pi_{gx}^2 \ 2p\pi_{gy}$	3		
F_2	$[Be_2] \ 2p\sigma_g^2 2p\pi_{ux}^2 2p\pi_{uy}^2 2p\pi_{gx}^2 \ 2p\pi_{gy}^2$	2	1.34	144

can come closer to ψ (Burrau) by expanding the basis set to include *p*-orbitals. Thus we might try

$$\varphi_2 = A'(1s_a + 1s_b) + A''(2p_{za} + 2p_{zb}) \qquad (4.3)$$

I must remind you that the electron is not jumping from one to the other of the four atomic orbitals. The whole idea is to get close to the true MO by adding known functions. The basis functions in the summation have no significance other than to indicate that we need them to "synthesize" a function that comes close to the accurate solution.

There are further refinements possible. As in the case of the helium atom we can adjust the nuclear charge. After all, the electron experiences the potential of two positive charges, each screened to some extent. The result of these modifications gives 2.71 eV for dissociation energy, very close to the experimental value.

4.2. Methodology of Quantum Calculations

The above brief foray into chemical bond calculations might leave you with the impression that we mix functions until we get the right answer, just like we mix paints to get the right color, but that would be misleading. There are principles and constraints that make quantum chemistry a predictive discipline rather than one that rationalizes what is already known. These are examined in the following section.

4.2.1. *The variation theorem*

According to this theorem (usually called a principle), the energy you calculate with any trial function will be greater than the true energy (see also Section 3.2.3). You can never go below the true energy, but you get close to it as you refine your calculations and the exact energy, if you happened to have the right function. (The principle is valid only for the lowest energy state of a given symmetry). Thus in the H_2^+ calculations, as you add more terms or modify functions based on chemical intuition, you approach the true value but never go below it. Hence quantum calculations on the lowest energy state proceed along the following lines:

(a) Find the wave function that gives the lowest energy.
(b) Use this function to calculate the properties of bonds and molecules.

Energy is used as a test of how "good" the wave functions are, but the real purpose of the calculations is to explain the chemical properties.

4.2.2. *Complete sets*

The three unit vectors (along the x, y and z axes), also called the *basis set*, form a complete set for Euclidean space in the sense that any line in that space is a weighted sum of the projections along these axes. Similarly, the mathematicians talk about "Hilbert space" which, crudely stated, has functions as bases. It turns out that eigenfunctions of a Hamiltonian form a complete set in Hilbert space. That means every function in that space is either one of the bases, or a linear combination of those bases. The wave function that corresponds to the exact energy of the lowest state is one of the axes. Any other function one could guess (with the correct Hamiltonian) for the lowest energy will have a component along that axis and give only an approximate value for the lowest energy state. As you refine your energy calculations, your trial function comes closer and closer to the lowest energy axis. Thus variation theorem is a consequence of the properties of Hilbert space.

My aim here is to give you an idea of the foundations of quantum calculations. Rigorous justification of anything I have said is beyond the scope of this discussion, but the following example illustrates the idea.

4.2.3. *Hydrogen atom in an electric field*

Consider a hydrogen atom in an electric field along the z-direction (Stark effect) whose Hilbert space is restricted to only two functions: ψ_1 and ψ_2. These are the eigenfunctions of the Hamiltonian for the hydrogen **atom in the field**. Figure 4.2 illustrates that "space."

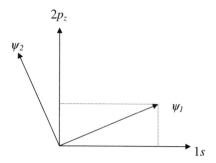

Fig. 4.2. The functions $1s$ and $2p_z$ are eigenfunctions of the Hamiltonian for the hydrogen atom. ψ_1 and ψ_2 are the eigenfunctions for the hydrogen atom in an electric field. With the hydrogen atom functions as a basis set we construct the functions in the field, illustrated here with just two basis functions, although in general, one chooses a large basis set for reliable calculations.

The eigenfunctions of a Hamiltonian have the following orthogonal property:

$$\int \psi_1^* \psi_2 dV = 0 \qquad (4.4)$$

In Fig. 4.2, these are shown perpendicular to each other to emphasize this property. Let us say that ψ_1 is the correct solution. We do not know what it looks like, but we have to choose a basis set to synthesize it. Here we are in the comfortable position of being able to visualize what is happening. We see that the natural basis functions are 1s (hydrogen atom before the field is on) and $2p_z$ since charge distortion is along the z-axis. Hence

$$\psi_1 = c_1(1s) + c_2(2p_z) \qquad (4.5)$$

We see that superposition of the two functions will give us the desired function once we determine the coefficients c_1 and c_2. By calculating energy as a function of c_1 and c_2 and minimizing the energy we obtain the numerical values of c_1 and c_2.

4.2.4. *Rules of superposition*

1. Mixing between basis orbitals becomes insignificant as the energy separation between them increases. The most pronounced mixing occurs between the orbitals whose energies are close.
2. The superposed wave function must conform to the symmetry of the Hamiltonian.
3. Probability must be conserved in linear combinations. In practice this means that if you superpose N basis functions, you must get N superposed functions to preserve the total probability.

Generally speaking, the accuracy of calculation improves as the number of proper basis functions included increases.

4.2.5. *Virtue of qualitative thinking*

One of the fascinating aspects of quantum theory is that you can gain valuable insight through qualitative reasoning alone.

When an atom is polarized, the centers of negative and positive charges are separated. Hence it acquires a dipole moment (μ) in the field. The acquired dipole moment is proportional to the electric field strength, E

$$\mu = \alpha E \tag{4.6}$$

In this equation $\alpha[\text{m}^3]$ is called the polarizability. The qualitative consideration of the H-atom in an electric field shows that α depends on mixing of orbitals.

The polarizability of Li is about 35 times that of H, and it is not difficult to see why. The outer electron in Li is in the $2s$ function which is almost degenerate with $2p_z$ function. Thus, the mixing of those two functions will be more pronounced than $1s$ and $2p_z$ in hydrogen, so we would expect Li to have a higher polarizability.

Exact calculations for polarizability require an extended basis set. However, as this example shows, we can often explain how properties vary from atom to atom or molecule to molecule by qualitative considerations alone. The wide use of quantum theory in chemistry depends to a large extent on its qualitative applications.

4.3. Homonuclear Diatomic Molecules

The first and second columns of Table 4.1 give the bond energies and bond lengths for diatomic molecules of the first and second-row elements. The data are from spectroscopic studies of gas phase molecules and ions at high temperatures.

The theory for understanding bonding in these molecules is a straightforward extension of the ideas we have just explored.

(1) We first construct MOs. For doing this we have two rules to guide us: (a) Since the two atoms are identical, the coefficient that goes with each atomic orbital must be the same. (b) The MOs have to be eigenfunctions of symmetry operators.

The MOs must have ± 1 for eigenvalues of the operators for inversion and reflections. Possible linear combinations are shown in Fig. 4.3. The shapes of MOs that result from atomic orbitals (AOs)

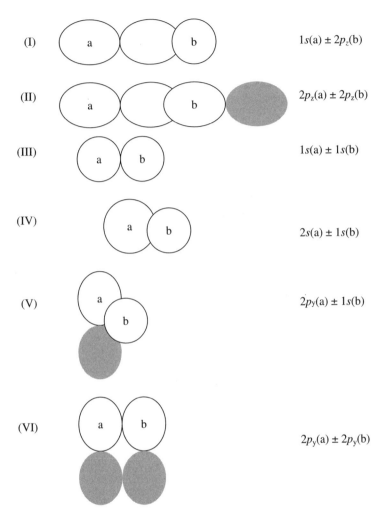

(I) $1s(a) \pm 2p_z(b)$

(II) $2p_z(a) \pm 2p_z(b)$

(III) $1s(a) \pm 1s(b)$

(IV) $2s(a) \pm 1s(b)$

(V) $2p_y(a) \pm 1s(b)$

(VI) $2p_y(a) \pm 2p_y(b)$

Fig. 4.3. Permitted and forbidden linear combinations. If a symmetry operator commutes with the Hamiltonian, the MOs have to be eigenfunctions of both operators. The symmetry operators for inversion and reflections in three Cartesian planes commute with the Hamiltonian for a homonuclear diatomic molecule. Only symmetry operators for reflections in the two planes containing internuclear axes commute with the Hamiltonian for a heteronuclear diatomic molecule. Only the linear combinations II, III, IV are allowed for homonuclear diatomic molecules, but every linear combination shown in the figure is allowed by symmetry for a heteronuclear diatomic molecule. However, V cannot contribute to bonding since the positive and negative overlaps cancel and IV can only be a minor player because of the energy gap rule.

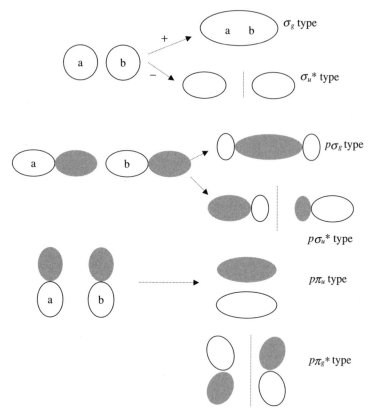

Fig. 4.4. MOs for homonuclear diatomic molecules formed from linear combination of AOs. These sketches are intended to give some indication of the shapes of the MOs and are not quantitatively correct. Note the bonding π orbitals have u symmetry and the antibonding π orbitals have g symmetry.

with $n = 1$ and 2, when these two rules are taken into account, are shown in Fig. 4.4.

The notation used to label the MOs was introduced by Mulliken. Subscripts g and u indicate whether the function is symmetric (eigenvalue 1) or antisymmetric (eigenvalue -1) for inversion. Letters σ (sigma) and π (pi) indicate s and p-like appearance when viewed along the internuclear axis. (In complex representations of these orbitals, σ and π refer to the angular momentum around the internuclear axis.)

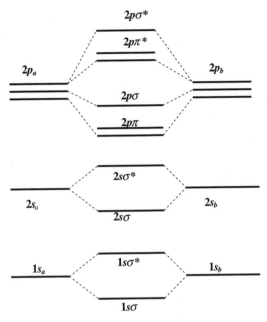

Fig. 4.5. Molecular orbital energies on a relative scale. The actual MO energies and their order are not fixed, but vary to some extent from molecule to molecule. The order chosen here is strictly for comparison. Stars indicate antibonding orbitals. Some antibonding orbitals have lower energy than bonding orbitals. This is because bonding or antibonding is with respect to the parent orbitals in the linear combination.

(2) We arrange these MOs in the order of increasing energy. Figure 4.5 shows the energy level diagram. The order in which the levels are arranged will be explained at the end of this section.

(3) Finally we "fill" the MOs according to the Pauli principle to get the lowest energy configurations for homonuclear diatomic molecules. The configurations thus generated are given in the second column of Table 4.1. We are now in a position to examine these results.

4.3.1. *Antibonding orbitals*

Notice that every bonding MO is paired with an antibonding MO in Fig. 4.5. This is truly a quantum effect for which there is no classical analog. Bonding and antibonding correspond to the constructive and destructive interference of waves. If we linearly combine the bonding and antibonding MOs, we will get the starting AOs.

The nature of bonding is determined by both bonding and antibonding MOs. Consider the case of N_2 and O_2 (Fig. 4.5 and Table 4.1). The extra electrons in O_2 go into the antibonding MOs, thus increasing the bond length and decreasing bond energy. As we can see from the table, the F_2 bonding is weakened further because there are four antibonding electrons and only two net bonding electrons.

Another interesting quantum effect is that when an electron is removed from N_2 the bond becomes weaker. The opposite happens when an electron is removed from O_2 because in N_2 the electron is ejected from a bonding orbital and in O_2, from an antibonding orbital. This sort of change cannot be explained except through the wave theory of matter. Note also that He_2^+ has one net bonding electron, just like H_2^+. As a result, the two molecular ions have comparable bond energies.

4.3.2. *Paramagnetism*

Among the diatomic molecules in Table 4.1, O_2 and B_2 are **paramagnetic** because of unpaired spins. (**Ferromagnetism** is from permanent orientation of spins; paramagnetism, from random orientation.) When you pour liquid oxygen (a light blue liquid) between the poles of a non-uniform magnet, the stream deflects toward the stronger pole. One of the success stories of MO theory was its ability to explain the paramagnetism of O_2 soon after the development of quantum theory.

4.3.3. *What is a molecule?*

Why do atoms form a bond? What is the nature of a chemical bond? A detailed discussion would take too long here, but the following examples will give you an idea of the sort of answers one might expect to such questions.

Be_2 was detected by laser ablation of Be atoms in a stream of helium gas cooled to about 70 K. The high-intensity laser beam was required to obtain sufficient atoms in the gas phase, while the immediate cooling was needed for the hot atoms to lose their energy and form a molecular bond. This example illustrates an important idea: ultimately molecule formation is a gift of the environment. Without the helium cooling, the Be atoms would not form a molecule. Any two atoms coming together in a vacuum will not stay together but bounce back, since energy has to be conserved. They would only settle into a molecular configuration if some of the excess energy of the atoms were removed by the surroundings or through emission of radiation.

Under high vacuum, a pair of helium atoms with an equilibrium distance of 100 times the average bond length showed the characteristics of a wave packet in single slit diffraction experiments. The binding energy was estimated to be approximately one tenth of non-bonded van der Waals interaction. Should we claim that the dihelium molecule exists? Or that the helium atom pair is a transient species that shows up under very restricted conditions?

Liquid oxygen is light blue in color even though oxygen gas is colorless. The color arises from two molecules of oxygen absorbing one photon. The frequency of absorption is twice the absorption frequency of a single molecule. We have to conclude from this observation that the binding energy between two O_2 molecules is negligible. The explanation given to the blue color of liquid oxygen is that the bi-molecular transition takes place during a short-duration collision, without formation of an O_4 molecule.

So what should be the minimum binding energy and lifetime before we accept a collection of atoms as a molecule? I am not interested in a dialectical discourse on the definition of a molecule. However, the above questions should make you realize that the formation and stability of a molecule ultimately depends on the surroundings and that under unusual circumstances we may find a collection of atoms that hang together and show some, but not all, of the properties we normally associate with molecules.

Today experiments are done at very high and very low temperatures and pressures, in electric and magnetic fields. These experiments make chemistry, which for a long time has been confined to ambient conditions, a richer discipline. They also demand, however, that we critically examine the words we use to explain the experiments.

4.3.4. *Ordering of energy levels*

Strictly speaking, we cannot trust a diagram like the one in Fig. 4.5 for a quantitative account of bonding in every molecule in Table 4.1. Potential energy is not the same for every molecule; nor are the forms of the MOs. For instance, $2s\sigma_g$ and $2p_z\sigma_g$ will further interact with varying degrees in different molecules to form MOs with a basis set of four functions. In Fig. 4.5, I have chosen the order that comes closest to the actual calculations, so that I can sketch a general theory.

4.4. Heteronuclear Diatomic Molecules

The MO theory follows the same rules as for homonuclear systems; the MOs are constructed from AOs as before. The theory, however, becomes more involved since we do not have orbitals with the same energy on neighboring atoms. There is no inversion symmetry in the heteronuclear diatomic molecules to guide us; hence the coefficients for AOs on the two atoms will not be the same.

Consider the example of LiH. (Note the formula is not expressed as HLi, even though the two symbols represent the same entity physically. Chemists knew from various experiments that the negative end of the molecule happened to be on the H atom.)

Figure 4.6 illustrates the relative energies of the low lying AOs of hydrogen and lithium. The rules that guide us in generating MOs are (1) the AOs must have the same symmetry for rotation around the internuclear axis and (2) the AOs closest in energy interact most. The electron $1s_{Li}$ feels the nuclear charge of three protons and consequently has much lower energy than the hydrogen $1s$ function, so it is not likely to interact with $1s_H$ appreciably. Of the three $2p$ functions of lithium, only one has the correct symmetry with

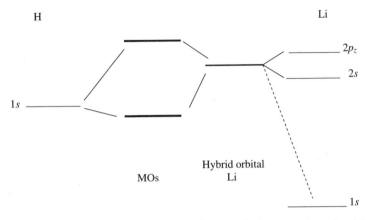

H   Li

Fig. 4.6. Bonding in LiH. The figure shows relative energies of orbitals in hydrogen and lithium. Since there is no center of symmetry, linear combinations do not have to satisfy the requirements of inversion or reflection in the plane perpendicular to the internuclear axis. Hence any one or more of the Li orbitals shown can interact with the hydrogen $1s$ orbital. However, the energy gap rule indicates that in the linear combination only $1s_H$, $2s_{Li}$ and $2p_{z,Li}$ will have significant coefficients. In the scenario shown here, the orbitals in Li hybridize (mix) in the presence of H and one of the hybrid orbitals combines with the $1s_H$ to form MOs.

respect to the internuclear axis. $2s_{Li}$ and $2p_{z,Li}$ are close in energy. This suggests that the bonding in LiH can be accounted for (1) by hybridizing the $2s_{Li}$ and $2p_{z,Li}$ functions and (2) making an MO from one of the hybrid orbitals and $1s_H$. This is illustrated in Fig. 4.6.

More sophisticated calculations take the $1s$, $2s$, $2p_z$ functions of Li and the $1s$ function of H as the basis set and vary in the nuclear charge for each AO. The bonding MO in one of those calculations is given by

$$\Psi = -0.131(1s_{Li}) + \{0.323(2s_{Li}) + 0.231(2p_{z,Li}) + 0.685(1s_H)\} \quad (4.7)$$

Notice that the three functions in the braces have the same phase (same sign). Both $2s_{Li}$ and $2p_{z,Li}$ orbitals contribute to bonding. However, $1s_{Li}$ also affects bonding. In other words, the hybridization on lithium involves all three AOs. Since the square of the wave function is probability density, it is clear that $1s_{Li}$ is a minor player: $(0.131^2 = 0.017)$.

The dipole moment of LiH, calculated with the above function, is in excellent agreement with the experiment.

4.5. The Valence Bond Theory

Immediately following the development of wave mechanics, two apparently rival theories of the chemical bond appeared. We have already discussed molecular orbital theory. The other, called valence bond theory, was formulated by Heitler and London. The wave function for the lowest energy state of H_2 in this theory is given by

$$\psi(\text{VB}) = \frac{1}{\sqrt{2}}[1s_a(1)1s_b(2) + 1s_a(2)1s_b(1)] \qquad (4.8)$$

The idea behind the theory is that, since electrons are indistinguishable, the configuration $1s_a(1)1s_b(2)$ is as likely as the configuration $1s_a(2)1s_b(1)$; hence both must be included in the wave function. The theory was an immediate success, unlike the MO theory, since it gave better values for the dissociation energy, and chemists were able to visualize the two configurations as two resonance structures. It soon became clear, however, that both theories needed refining and ultimately, they led to the same results. Figure 4.7 gives the methodological differences between the VB and MO theories. As you can see from this figure, MO theory first superposes basis functions (usually atomic orbitals) and then takes products, while the VB theory starts with the products of AOs and then superposes the configurations. Many quantum phenomena involve interference effects. Since the MO theory starts with superposition it is in a better position to give a qualitative explanation of the quantum properties of matter which arise from superposition. Hence it is more frequently relied upon now, even though VB theory continues to find applications in molecular structure calculations.

4.6. Spectra of Diatomic Molecules

Transitions between electronic states are observed in the visible and ultraviolet region, accompanied by vibrational and rotational

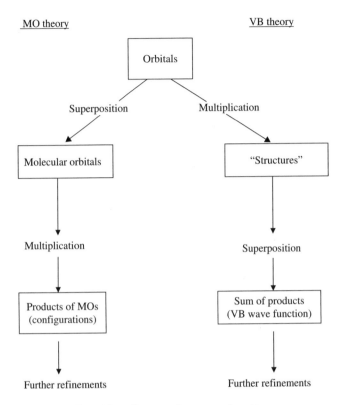

Fig. 4.7.   See text for an explanation.

transitions. This means that information on bond lengths can also be obtained from a high-resolution electronic spectrum. Other information obtained from electronic spectra includes dissociation energies, bond lengths and reactivity of excited states.

4.6.1. *Potential energy curves: Morse potential*

Figure 4.8 shows potential energy curves for two electronic states, X and A, of a diatomic molecule. The first thing we should notice is that the curves X and A are not what we would normally expect for a harmonic oscillator, which cannot dissociate. However, the diatomic molecule can, if the energy is right. The harmonic oscillator approximation is only reasonable at the bottom of the curves.

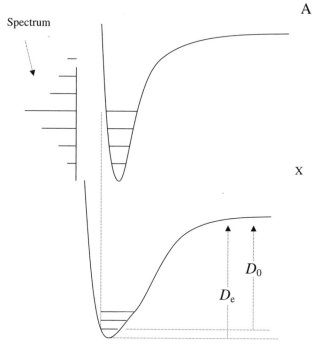

Fig. 4.8. Morse curves for lower and upper electronic terms. Intensity of vibrational bands is determined by the overlap of the vibrational wave functions as shown in Fig. 4.9. The expected spectrum is indicted here by lines to keep the diagram simple. In the actual spectrum these lines are broad bands because of the accompanying rotational transitions.

We need a more realistic potential energy curve for understanding electronic spectra of diatomic molecules.

A frequently used potential energy function is due to Morse. The equation for the Morse potential curve is

$$E(\text{pot}) = D_e(1 - e^{-\beta(R-R_e)})^2 \tag{4.9}$$

In this equation β is a constant, D_e is the dissociation energy measured from the bottom of the curve and R_e is the equilibrium distance.

When $R = R_e$, we have $E(\text{pot}, R_e) = 0$, the bottom of the well. When $R > R_e$, $E(\text{pot}, R) = D_e$, the dissociation energy measured from the bottom of the well. When $R < R_e$,

$$E(\text{pot}) = D_e(1 - e^{\beta R_e})^2 \approx D_e e^{2\beta R_e}$$

which results in a large positive number (depending on β) showing repulsion between atoms as they come close beyond the equilibrium distance.

Morse potential, rather than harmonic oscillator potential, must be used in the Schrödinger equation to get the vibrational energies of an anharmonic oscillator.

4.6.2. *Vibrational structure of electronic transitions*

We have indicated a few vibrational levels in X and A in Fig. 4.8. There are usually many more as between the vibrational states there are rotational states, which are not shown. Figure 4.8 also indicates the expected spectrum. I have assumed that only the $v'' = 0$ level is populated, which is usually the case at ambient temperature (spectroscopists use double prime for the lower state and single prime for the upper state). Absorption is a series of bands from $v'' = 0$ to $v' = 0, 1, 2, \ldots$ There is no vibrational selection rule for transitions between electronic states.

The relative intensities of vibrational peaks are determined by the Franck–Condon principle. According to Franck, the nuclei may be considered stationary while the electrons undergo transitions. Thus transitions take place vertically from whatever configurations the two atoms happen to be in. Condon gave a more refined and more useful quantum mechanical explanation for the intensities. According to Condon, the intensity of a vibrational band is proportional to the overlap of the vibrational wave functions. Thus a transition from v'' to v' will have intensity proportional to

$$\int \psi_{v''}\psi_{v'}dR$$

This is shown in Fig. 4.9.

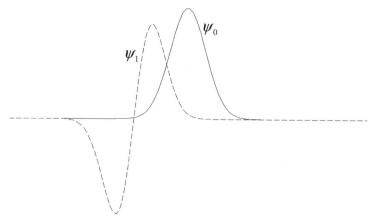

Fig. 4.9. Intensity of a vibrational band in an electronic transition depends on the overlap integral, $\int \psi_0 \psi_1 \, dr$. In this integral, r is the internuclear separation; one vibrational wave function belongs to the lower electronic state and the other to the upper electronic state.

4.6.3. *Dissociation and predissociation*

Figure 4.10 shows potential energy curves for two excited states, B and C. The lowest energy state, X, is indicated by a horizontal line and harmonic and the wave function for the $v'' = 0$ vibrational level. A molecule excited to C dissociates, since it is not a bound state. Hence there are no quantized vibrational or rotational energies in C and X $\rightarrow$ C absorption is a continuum. In fact, the absorption spectrum from the $v'' = 0$ level looks like a distorted image of the $v'' = 0$ wave function as shown in Fig. 4.10.

In the region marked predissociation, the C and B curves cross. The rotational-vibration structure of the X $\rightarrow$ B transition gets blurred in this region because molecules in the B state can go over into the dissociative continuum. The spectrum beyond this point again becomes discrete. This region represents **predissociation**, in contrast to true dissociation which happens at higher energies.

4.6.4. *Dissociation energy*

From the vibrational energies in an electronic state we can estimate the dissociation energy. Due to anharmonicity, the separation

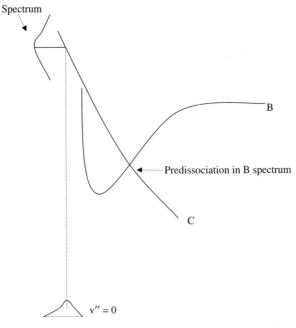

Fig. 4.10. Dissociation and predissociation. There is no quantization in the C term. The spectrum X → C is simply a reflection of the $v'' = 0$ function of the potential wall of C. X → B spectrum is not shown in the figure. The spectrum becomes blurred at the crossing point of B and C due to the coupling of these two terms. Above and below this point the spectrum shows discrete vibrational bands.

between successive vibrational levels decreases. The sum of all these separations must be equal to the dissociation energy. However, data are usually available for a limited number of vibrational energies, so the differences between available energies are used to extrapolate to the point where the difference vanishes. The dissociation energy thus obtained is D_o, since it is measured from the 0th vibrational level. D_e in the Morse function is related to this quantity by

$$D_e = D_o + 1/2h\nu \qquad (4.10)$$

4.6.5. *What are the electrons doing?*

One of the common misunderstandings in electronic spectra is to assume that a particular electron is jumping from one orbital to

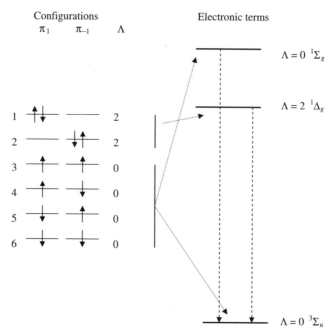

Fig. 4.11. The six configurations (products of two orbitals), shown on the left, are degenerate, but have been drawn as if they are not, to make the diagram simple. The two orbitals differ in their angular momentum around the internuclear axis, which is shown as a subscript. When electron repulsion is included in the energy calculations, the configurations group and split into three electronic terms. Electronic transitions are between terms and *not* between orbitals. The transitions shown above are observed in near infrared and far ultraviolet. Capital Greek letters are used to denote electronic terms. Besides the subscripts u and g, you will see $+$ and $-$ as superscripts, to represent the symmetry properties of terms.

the other. We have to go beyond the orbital scheme to understand spectroscopic transitions. Let us see what really happens with O_2 as an example.

The potential energies shown in Figs. 4.8 and 4.10 do not correspond to orbital energies. The orbital picture, as we have seen in the case of atoms, does not take into account electron repulsion explicitly. When you include electron repulsion, you get linear combinations of configurations called **electronic terms**. Fortunately, we do not have to compute the energies of these terms. All that is needed is information of the angular momentum of each term, which is obtained by simply adding the angular momenta of electrons.

In the orbital diagram for O_2 we placed one electron in each π orbital. The symbol for orbital angular momentum around the internuclear axis is λ. In the complex representation, the π orbitals have $\lambda = 1$ and $\lambda = -1$. When we couple them we get for the combined angular momentum, Λ, values 2, 0, 0 and -2 as shown in Fig. 4.11. $\Lambda = 2$ and -2 must belong to one term. It is called the Δ term. Since we have to put both electrons in the same π orbital, the spins have to be anti-parallel giving a singlet spin term. The two terms with $\Lambda = 0$ correspond to $^1\Sigma$ and $^3\Sigma$ as shown in Fig. 4.11.

This chapter is very selective, since bonding in molecules covers a very large area. Bonding and spectra of diatomic molecules are the essential foundations for understanding polyatomic molecules. It is interesting to note that we can make sense of bonding from a qualitative consideration of symmetry and the relative energies of hydrogen-like orbitals.

 Related material on the disc.

With CD-ROM

Part B

Chemical Thermodynamics

I will start with a list of chemical and physical changes that affect our lives:

1. $O_3(g) + O(g) \leftrightarrow 2O_2(g)$
   (Gas phase atmospheric reaction leading to ozone depletion)
2. $N_2(g) + 3H_2(g) \leftrightarrow 2NH_3(g)$
   (Haber process for manufacture of ammonia, used in fertilizers and explosives)
3. $CaCO_3(aragonite, s) \leftrightarrow CaCO_3(calcite, s)$
   (Equilibrium between two solid phases of interest to geochemists)
4. $H_2O(l) \leftrightarrow H_2O(g)$
   (Phase transition that provides steam power)
5. $C_6H_{12}O_6(aq) + 6O_2(aq) \leftrightarrow 6CO_2(g) + 6H_2O(aq)$
   (Metabolic reaction)

These chemical transformations were chosen from different areas of science to call your attention to the breadth of the subject we are about to explore. Each is a chemical reaction. For each of the reactions the relevant questions are (1) what is the abundance of various species when they are at equilibrium at a particular temperature and pressure (environmental variables), and (2) how does the equilibrium change when we change the environmental variables?

In thermodynamics, a system is separated from the surroundings by a boundary. Since each of the reactions above is carried out in an enclosed vessel, the species in each reaction make up a *system*. Everything else is *surroundings*. If we are interested in studying a sample of gas without any chemical changes, the gas is the system. The distinction between system and surroundings is crucial to thermodynamics since the theory depends on distinguishing changes within the system from the processes (effects of interactions) between the system and the surroundings.

Equilibrium and reversibility

A system that does not change with time is said to be in equilibrium. This is true but there is more to it. A system in equilibrium macroscopically (its pressure, temperature and chemical composition stay fixed) is continuously changing on the microscopic level. Thus when a liquid is in equilibrium with its vapor, pressure and temperature do not vary but the molecules are continuously exchanging their positions within and between the phases. To say thermodynamic equilibrium is a dynamic equilibrium is not an oxymoron.

Microscopic reversibility (products and reactants continuously transforming into one another) is a property of thermodynamic equilibrium. For reasons that will be explained in Section 6.2, the formal development of thermodynamics introduced the idea of the reversible **process**, which should not be confused with reversible **changes** within the system. A reversible process is a change imposed on the system while it continues to be in equilibrium with the surroundings. Such a process cannot be achieved in practice. The closest we can get is a quasi-static process where the system is perturbed as little as possible. For example, to investigate heat absorption by the system, we restrict the temperature change to the minimum value that can be measured reliably.

Chapter 5

Entropy and Equilibrium

Equilibrium is determined by **energy** and **entropy**. The laws which constrain the direction and extent of change are:

(1) Energy gained (or lost) by a system (object of interest) is equal to energy lost (or gained) by the surroundings with which the system interacts. Energy of an **isolated system** remains constant, since it does not interact with the surroundings.

(2) Entropy of an isolated system reaches a maximum at equilibrium. If the system and part of the surroundings are isolated, entropy of the combined system reaches a maximum.

These are known as the first and second laws of thermodynamics. In the next four chapters we will explore their important role in chemistry.

The word "entropy" has permeated so many widely different areas that it now has an aura of mystery, surrounded by a fog of misunderstanding. Yet entropy is a very simple concept, if we stick to classical and statistical equilibrium thermodynamics. Unfortunately, it becomes opaque and confusing if incorrect models or inappropriate words are used to describe it. Since clear understanding of entropy *as it is related to equilibrium* is essential, let us consider a simple thought experiment that illustrates the nature of entropy.

Say we have four diatomic molecules, A, B, C and D, and each can have 0, 1 or 2 $h\nu$ units with a total energy (harmonic oscillator energies). We will start the experiment with all the molecules in the lowest possible energy state as shown below.

$$E = 2 \quad \underline{\hspace{3cm}}$$
$$E = 1 \quad \underline{\hspace{3cm}}$$
$$E = 0 \quad \underline{\text{A} \ \text{B} \ \text{C} \ \text{D}}$$

Now let us selectively excite a molecule, say A, to the $E = 2$ level and isolate the four molecules so that energy cannot be lost to the surroundings, but has to remain with the four molecules. Column 2 in Table 5.1 shows the arrangement immediately after excitation.

Will the energy remain for all time with molecule A? The answer is no. It will be shuffled among the four molecules. We cannot restrict it from being distributed in every possible way provided the total energy remains exactly two units.

There are ten different ways of distributing energy among the four molecules without violating this rule (among them is the one to which A is selectively excited). The ten **microstates** are shown in the shaded region of the table. We have to assume that each of the microstates is equally probable. In ten million measurements, for instance, we expect to find each microstate in the shaded area a million times. Why do we have to assume this? Suppose we do not. Then we would be *assuming* that nature is biased, preferring some arrangements over others without reason.

Table 5.1. Ten different microstates available for a system of four molecules with three equally-spaced quantum energy states.

Energy states	Microstates at equilibrium									
E = 2	A	B	C	D						
E = 1					AB	AC	AD	BC	BD	CD
E = 0	BCD	CDA	DAB	CAB	CD	BD	BC	AD	AC	AB

This example illustrates two important principles:

(1) Total energy of an isolated system is constant (the first law of thermodynamics). In this example, energy must be two units throughout all microscopic changes.
(2) At equilibrium, energy becomes distributed among all available microstates (a consequence of the second law of thermodynamics). In the above example, energy, which initially "belongs to" molecule A, is spread out over ten different arrangements.

We can state the results of our thought experiment thus. The system is not at equilibrium when one molecule (A in our example) has all the energy. The system reaches a **dynamic equilibrium** when energy is spread among all possible energy states (the ten microstates in the shaded area in our example). The equilibrium is dynamic since energy shuffling does not cease when equilibrium is achieved.

It is important to distinguish microstates from quantum states. A microstate spans all molecules in the container; there is one quantum state for each molecule in a microstate. In the example we have considered, each microstate has the same energy. A quantum state is an energy state available for each molecule. Quantum states differ in their energy. More than one molecule may have the same quantum state (we usually say that more than one molecule can occupy a quantum state). Distribution among quantum states needs not be the same as distribution among microstates.

5.1. Entropy: A Measure of Energy Shuffling

Max Planck (1900), extending the pioneering work of Ludwig Boltzmann (1872), defined entropy (S) by the equation

$$S = k_B \ln \Omega \qquad (5.1)$$

where Ω is the number of microstates. This equation is known as the Boltzmann entropy equation. The constant, k_B, is called the Boltzmann constant. It is exactly equal to the gas constant divided by the Avogadro number

$$k_B = \frac{R}{N_{Avo}} = 1.380658 \times 10^{-23} \, \text{JK}^{-1} \qquad (5.2)$$

In our example, $\Omega = 1$ initially and $\Omega = 10$ at equilibrium. Initially entropy is zero ($\ln 1 = 0$) and at equilibrium it is $2.3\,k_B$, which is the maximum possible value under these conditions.

There is another way of stating the results of our thought experiment. *An isolated system comes to dynamic equilibrium when its entropy reaches a maximum for a given amount of energy.*

By defining entropy as a logarithmic function of the number of microstates, we made it an additive property. Consider two identical systems, I and II, each like the one discussed above. Each system has ten microstates. Since the behavior of I is not restricted by II, any microstate of I may appear in conjunction with any microstate of II. Thus the total number of microstates is a product of the microstates of each system

$$\Omega(I + II) = \Omega(I)\Omega(II) \qquad (5.3)$$

and there are a hundred microstates for the combined system.

Entropy, as defined by the Boltzmann equation, is a sum

$$S(I + II) = k \ln \Omega(I + II) = k \ln \Omega(I) + k \ln \Omega(II)$$
$$= 4.6\,k$$

Since entropy of a collection of systems can be added to get total entropy, it is called an **extensive property**. The advantages of working with these will become clear later. Temperature and pressure are **intensive properties**, since subdividing the system does not alter their values. The distinction between extensive and intensive properties is important in thermodynamic theory. A system is in equilibrium only when the intensive properties are uniform throughout the system. Thus if you subdivide a system into N parts, each part will have the same value for intensive properties.

The Boltzmann constant, k_B, makes the definition of entropy consistent with the earlier research of Clausius (1862) in classical thermodynamics.

5.2. Chemical and Physical Changes

We will now see how entropy, as a measure of distribution of energy, is the invisible hand behind many processes.

5.2.1. *Free expansion of a gas*

If a system of gas expands from a high to a low pressure region, the entropy increases. To see why, consider a sample of gas initially confined to the left bulb in Fig. 5.1. When the partition between the two bulbs is removed, the gas expands to fill the whole volume. This experiment can be done at low enough pressure to avoid the effects of attraction or repulsion between molecules. So the only cause of expansion is the entropy change, which can be estimated.

We do not know the absolute number of physical sites available for a molecule. The number, however, must be proportional to the available volume. Therefore

$$\Omega(\text{initial}) = cV(\text{initial})$$
$$\Omega(\text{final}) = cV(\text{final})$$

for *each* molecule. Here c is a constant. Hence entropy change per molecule is given by

$$\Delta S(\text{molecule}) = k \ln \frac{\Omega(\text{final})}{\Omega(\text{initial})} = k \ln \frac{V(\text{final})}{V(\text{initial})}$$

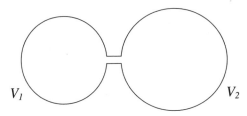

Fig. 5.1. The expansion of an ideal gas is solely the result of entropy increase. There is no energy change since there is no interaction between molecules. When a gas initially confined to volume V_1 expands to fill both bulbs, its entropy change is given by $\Delta S = nR \ln \frac{V_1 + V_2}{V_1}$.

Since entropy is an extensive property, entropy change for N molecules is given by

$$\Delta S(N \text{ molecules}) = Nk \ln \frac{V(\text{final})}{V(\text{initial})} \qquad (5.4)$$

If instead of the number of molecules, moles are used in the calculation, the above expression takes the form

$$\Delta S = nR \ln \frac{V(\text{final})}{V(\text{initial})} \qquad (5.5)$$

where n is the number of moles.

5.2.2. *Formation of solutions*

Gases are observed to mix spontaneously. Since intermolecular energy is negligible except at high pressures, the driving force for mixing is entropy.

Let's say that 0.050 mol of O_2 in a 1.0 liter vessel and 0.150 mol of N_2 in 3.0 liter vessel are allowed to expand into a 4.0 liter vessel at constant temperature. The problem is the same as before except we have two gases expanding and mixing. Hence

$$\Delta S(\text{mixing}) = \Delta S(O_2) + \Delta S(N_2)$$
$$= R[0.050 \ln(4/1) + 0.150 \ln(4/3)]$$

A convenient measure of concentration in solution chemistry is mole fraction, X. Mole fraction of compound A is simply n_A/n, the ratio of number of moles of A to the total number of moles. For gases it is the same as the ratio of partial pressure to total pressure. From the above data, it is readily seen that the mole fractions of O_2 and N_2 are $(1/4)$ and $(3/4)$ respectively. Hence we will write the above expression as

$$\Delta S(\text{mixing}) = -R[0.050 \ln(1/4) + 0.150 \ln(3/4)]$$
$$-R[n(O_2) \ln X(O_2) + n(N_2) \ln X(N_2)] \qquad (5.6)$$

Extension of the above formula for mixing of several chemicals gives

$$\Delta S(\text{mixing}) = -R \sum n_i \ln X_i \qquad (5.7)$$

where X_i are the mole fractions. For a mole of a solution,

$$\Delta S(\text{mole of solution}) = -R \sum X_i \ln X_i \qquad (5.8)$$

These formulas apply to gases, liquids and solids when the components form ideal solutions, that is when (a) there is no attraction or repulsion between molecules or (b) when the interaction between like molecules is the same as the interaction between unlike molecules. We will be using these formulas again in Chapters 8 and 9 to derive a few important relations.

5.2.3. *Chemical equilibrium*

Consider the following reaction between two diatomic molecules:

$$A_2 + B_2 = 2\,AB$$

To illustrate the role of entropy, we will assume that the reaction involves no energy change (I will justify this assumption soon). Since there is no energy change, we may conclude that the molecules have no "incentive" to react. However, that is incorrect as entropy changes can also drive the reaction. We can estimate entropy changes from the following argument. Suppose we start with equal numbers of A_2 and B_2 molecules. An $A_2 \ldots B_2$ pair has equal probability of producing two AB molecules or one A_2 and one B_2 molecule (Fig. 5.2). This means that there will be two AB molecules for every single A_2 and B_2 molecule when equilibrium is reached.

The forward rate of reaction is

$$\text{Rate}(A_2 + B_2 \rightarrow 2AB) = k_f[A_2][B_2]$$

and the reverse rate is

$$\text{Rate}(2AB \rightarrow A_2 + B_2) = k_r[AB]^2$$

where k_f and k_r are the rate constants. At equilibrium the two rates must be equal (microscopic reversibility). Hence the equilibrium

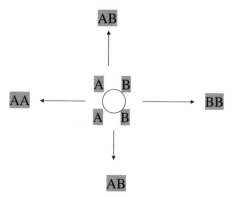

Fig. 5.2. A collision complex of two AB molecules, A_2B_2, breaks into (a) either two AB molecules or (b) one AA and one BB molecule.

constant, K, is

$$K = \frac{k_f}{k_r} = \frac{[AB]^2}{[A_2][B_2]} = 4$$

There is a real system that comes close to the above paradigm. This is the reaction between hydrogen and deuterium molecules

$$H_2 + D_2 = 2HD$$

for which the equilibrium constant is very nearly four at high temperature. There is, however, a small energy difference between the three species, and at low temperature, the equilibrium constant is less than four. We will discuss how it is possible to separate the roles of energy and entropy in Chapter 8.

5.2.4. *Molecule formation*

Two hydrogen atoms form an H_2 molecule because the molecule has lower energy. Right? The statement is only partially true. Suppose you bring two atoms together in a vacuum and somehow isolate them from the rest of the universe. What happens to their energy when they come close enough to form a molecule? They will have the same energy as two atoms (energy is conserved) and cannot stay together.

Yet we know that molecules form from atoms. The invisible hand here is entropy. The hydrogen molecule has discrete vibrational and rotational energy states. The surroundings (other particles, the vessel and everything in contact with the vessel) have far more energy states, densely packed. Thus, if the energy for the formation of the molecule is transferred to the surroundings, the number of states available for energy distribution and hence entropy, increases. Without that net entropy increase, there will be no molecule formation.

The reach of chemistry has expanded in recent years. Today we study chemicals at near absolute zero temperature, in high vacuum, at supersonic speeds and in interstellar space. The surroundings for these studies are not the same as those in an ordinary chemistry laboratory. Hence entropy changes could be different, leading to different and unexpected chemical processes.

5.3. Entropy: Wider Context

The word "entropy" originally confined to thermodynamics, has acquired many different meanings, some consistent with thermodynamic theory and others, less so.

5.3.1. *Entropy and disorder*

It is often said that entropy is a measure of disorder. What is disorder? Suppose you see a tutor's office with books stacked all over the place. You are right in concluding that it is an example of disorder. If the tutor leaves things alone (say, goes on a vacation) the books continue to be where they were left. So the whole arrangement, disordered, represents just one state. Its entropy is zero. On the other hand, many states are sampled if a compulsive tutor keeps on shuffling the books into a different order and entropy increases. Thus a person trying to create order might increase entropy more than one who leaves things disordered.

Take the familiar example of a pack of cards. They are in one "microstate" whatever the order might be and so entropy vanishes. It increases only when they are shuffled repeatedly, when all microstates are sampled.

From these two examples it should be clear that connecting the words disorder and entropy could be misleading. If in thermodynamics you mean by disorder a static arrangement, however ugly it might be, you are wrong.

What do we then mean by saying a crystal is more ordered than a liquid? It is true that molecules in a crystal are arranged in an orderly fashion compared to those in the liquid. So a crystal has a better order, in the **structural sense**, than a liquid. As a consequence, molecules have less freedom in a crystal (have fewer states to sample). One could say that entropy is a measure of freedom. I will not push that point and confuse the issue. It is the reduction in the number of states available that lowers the entropy and creates structural order, and there is no reason to avoid using the word "order" to indicate structural features as long as we understand that entropy is ultimately related to the number of energy states sampled by the system.

5.3.2. *Entropy and heat death*

The scenario for heat death is, put simply, that since entropy continues to increase as energy spreads over available states, ultimately energy will be uniformly distributed over all regions of space. No part of the space can amass energy and there will be one value for universal temperature. Hence life and all natural process that depend on energy flow cease.

The idea of heat death does, however, depend on some questionable assumptions. Clausius first showed that entropy of an isolated system increases spontaneously. To extend this idea to systems in contact with a heat bath, Clausius argued that the heat bath plus the system forms an isolated system. The idea is then further extended to the whole universe. His conjuncture may have been correct to a point, but it is questionable whether we can scale the process up from a heat bath to the universe. Clausius was reasoning in pre-quantum and pre-relativity days. According to quantum theory, there will always be a minimum kinetic energy, and according to cosmologists we still don't know the nature of 96% of the energy in the universe. Clausius was also unaware of the theory of fluctuations developed by Einstein in 1905. Hence Clausius' claim has to be reexamined. Whether Clausius is right or wrong, you can be assured that the universe is safe from heat death for the foreseeable future.

5.3.3. *Entropy and life*

You may have heard the following claims about life and thermodynamics:

1. Life violates the second law.
2. Lifeforms are highly ordered.

It is surprising that the first statement still continues to find favor with some people. If life violates the second law, so does the formation of the H_2 molecule, condensation of water vapor into clouds and various other phenomena. There is no violation of the second law as long as a decrease in entropy in one region is compensated for by its increase elsewhere. The entropy decrease in living organisms is invariably coupled with a greater entropy increase in the environment.

The second statement is ambiguous. What does one mean by "order"? Let us assume that one means by order a decrease in (a) entropy or (b) the number of states available for the system. Thus defined, we can get a number for order.

(a) In rough numbers the energy for basal metabolism per person (a measure of minimum energy needed to continue living) is 6×10^3 kJ/day. A person dissipates this as heat to the surroundings at 310 K (body temperature). So entropy decrease in the body is -13 kJ K^{-1} day^{-1}. This is the entropy decrease when 2.4 l of water vapor condenses at 298 K. It seems that there is no reason to congratulate ourselves by saying we are highly ordered. Incidentally, on the average men dissipate more energy than women. Are the men more ordered?

(b) The human body is estimated to have 6×10^{13} eukaryotic cells. These cells have to be packed in a particular pattern for a person to function. There are $6 \times 10^{13}!$ ways of arranging these cells and very few of them correspond to the structure we want. Let us say only one arrangement works. Then entropy decrease per cell due to formation of a human body from cells is

$$\Delta S = k_B \ln \frac{1}{6 \times 10^{13}!} \approx -2.5 \times 10^{-8} \mathrm{JK}^{-1}$$

almost an insignificant number. Even if we start with molecules instead of cells, we get a number in the above range. I will let you draw your own conclusions from this.

The numbers in (a) and (b) do not agree. That is because there is no specific meaning for the word "order" in this context.

5.4. Entropy and Heat

The entropy concept came into circulation before Boltzmann developed the statistical basis for entropy. In classical (or phenomenological) thermodynamics, entropy is connected to heat. It is common experience that heat will "flow" from a hotter to a colder body. Both heat flow when there is a temperature gradient, and increase in entropy are spontaneous processes which do not require energy input to the system, and there should be a connection between them. Suppose we bring into contact two blocks (A and B) of the same solid material at slightly different temperatures and allow a miniscule amount of heat, δq, to flow between them.

$$|\delta q_A| = |\delta q_B| = \delta q$$

However,

$$\frac{\delta q}{T_A} + \frac{\delta q}{T_B} > 0 \tag{5.9}$$

as long as the two temperatures differ. From the above two equations, it would appear that spontaneous change is related to $\delta q/T$ and not to δq alone.

Clausius showed that entropy increase in an irreversible process is greater than heat measured.

$$\delta S > \frac{\delta q_{irr}}{T} \tag{5.10}$$

Clausius also defined entropy in a reversible process by the relation

$$\boxed{\delta S \equiv \frac{\delta q_{rev}}{T}} \tag{5.11}$$

where the subscript "*rev*" stands for a reversible heat transfer between system and surroundings. According to some authors, this equation is a statement of the second law of thermodynamics. Since a reversible process is an idealization that can never be achieved (the closest we can get is a quasi-static process), you are likely to wonder how Eq. (5.11) can ever be used. As we will see in the next chapter, this equation allows us to develop the theory further and provides a connection between entropy and quantities that can be measured.

Particularly important ideas about entropy in this chapter are the relation of entropy to heat and to mixing of chemicals. Both these themes will appear in subsequent chapters.

 Related material on the disc.

With CD-ROM

Chapter 6

The Fundamental Equation
of Thermodynamics

It is useful to remind ourselves here that in thermodynamics a **system** (a bounded region with materials) is what interests us and **surroundings** are what interact with the system. Thus if you want to study the vapor pressure of a liquid, vapor and liquid form the system; the heat bath is the surroundings. Simple as this concept is, it is vitally important, because to follow thermodynamic reasoning we must clearly distinguish between **property change** within the system and the **processes** by which those changes are effected during a system's interaction with the surroundings.

6.1. The Fundamental Equation

According to the first law of thermodynamics, energy change, dU, in a system must be equal to the sum of heat and work exchanged with the surroundings.

$$dU = \delta q + \delta w \tag{6.1}$$

The formula for work is

$$\delta w = -PdV \qquad (6.2)$$

This formula will be discussed in more detail in Section 6.5.3.

The Clausius definition of entropy (Eq. (5.11)) is:

$$\delta q_{rev} = TdS \qquad (6.3)$$

From the three equations above we get

$$\boxed{dU = TdS - PdV} \qquad (6.4)$$

Equation (6.4) is the confluence of empirical and theoretical ideas in a mathematical form. The empirical arguments are that heat and work are forms of energy in **transition**, and that energy is neither created nor destroyed. Thus energy lost (or gained) by the system is energy gained (or lost) by the surroundings through the transfer of heat and work. The theoretical ideas are that energy is a **function of state** (i.e. it only depends on the physical state of the system as measured by P, V and T values, and not on the previous history of the system). Energy change from one state of the system to another is independent of how that change happens even though the relative quantities of heat and work vary depending on the path between the two states.

Since the first term in Eq. (6.4) is the second law and the second term is the definition of work, Eq. (6.4) looks like the **combined first and second laws**. This is certainly true, but there is a profound difference between this equation and Eq. (6.1).

Equation (6.4) relates energy changes to *the variables of the system*. Equation (6.1), on the other hand, relates energy changes to *processes*. Much of the confusion in understanding thermodynamics comes from not properly distinguishing between the two.

Between Eq. (6.4) and (6.1) lies an important chapter in the history of thermodynamics. In the application of scientific theories we can often ignore the historical development of the subject, and just accept derivations from theory as they arise. In thermodynamics, however, the ghosts of the past linger on and often lead to misunderstandings. A brief account of these developments should help you to be wary of these ghosts.

6.2. A Brief History

James Watt's name is synonymous with the steam engine because his work made it a practical device. Model steam engines were known before his arrival on the scene, but they could not be used in the field. Watt realized that the problem was that heating and cooling were being done in the same chamber and a large portion of heat was going into raising the temperature of the chamber before the power stroke. He solved the problem by connecting the steam chamber to a condenser. He also introduced the indicator diagram shown in Fig. 6.1 for estimating the actual work. This led to depicting the performance of the engines by cycles on a $P-V$ plane. The aim of the engineers at that time was getting maximum work from a given amount of heat. They knew heat engines used only a portion of the available heat, but they did not understand why.

As improvements in engine design were taking place, the need to know the maximum *theoretically* possible efficiency for conversion of

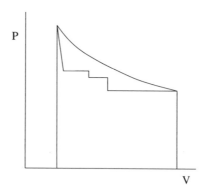

Fig. 6.1. Indicator diagram for work. The top curve represents reversible expansion which gives the maximum possible theoretical work (area under this curve). The step function below represents a possible expansion and the area under these steps is the actual work done by the engine. Reversible work is calculated from the formula $w = -\int P\,dV$. Actual work has to be estimated from the indicator diagram.

heat to work became clear. Obviously that would not give a blueprint to the engineer, but it would set the upper limit for efficiency of real engines.

By efficiency, ε, we mean work done by the engine per unit of heat absorbed in a cycle

$$\varepsilon = \frac{-w}{q_1} \qquad (6.5)$$

The numerator is work done by the engine; the denominator is heat absorbed at high temperature. The signs are chosen to indicate gain $(+)$ and loss $(-)$ for the engine.

The formula for maximum efficiency was developed by Sadi Carnot in 1824. He introduced the idea of **reversible work** (Fig. 6.1), not for practical reasons, but to have a measure of the maximum possible work. Carnot devised an idealized (hypothetical) cycle (Fig. 6.2), where every movement happens under reversible conditions, and proved that this cycle gives the maximum possible efficiency. His formula for maximum efficiency, ε, is

$$\varepsilon = 1 + \frac{q_2}{q_1} \qquad (6.6)$$

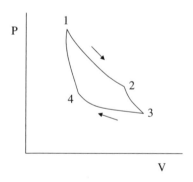

Fig. 6.2.   The Carnot cycle. $1 \rightarrow 2$ and $3 \rightarrow 4$ are constant temperature strokes (isothermal work), $2 \rightarrow 3$ and $4 \rightarrow 1$ are adiabatic strokes, where no heat is allowed to flow in or out of the system. Carnot proved that the efficiency of such a cycle, with each step carried out under reversible conditions, will have the maximum efficiency. This is a theoretical cycle that gives the upper limit for the amount of work we can extract from a given quantity of heat. The Carnot cycle, although impractical, provided the concepts essential for the theoretical development of thermodynamics.

where q_1 and q_2 are heat exchanges at high and low temperatures (q_2 is a negative quantity since the system gives out heat at low temperature).

That energy is conserved was not known at the time Carnot was developing his theory. He assumed, rather, that heat is conserved and after a lengthy argument got the above answer. Suppose he knew that energy cannot be created or destroyed, and must be the same before and after the cycle. Then he would have realized that $-w = q_1 + q_2$, and Eq. (6.6) follows directly from Eq. (6.5). So what was the fuss about?

Perhaps the most significant result to come out of Carnot's research was that every machine (irrespective of whatever the working substance might be) operating between the same two temperatures under reversible conditions must have the same efficiency. The proof is quite simple. Suppose machine A is more efficient than machine B. We couple them so that machine B acts as a refrigerator and make the heat absorbed by A at high temperature exactly equal to the heat rejected by B. Since A is more efficient, heat rejected by it at low temperature has to be less than heat absorbed by B. Besides, by our original assumption, more work is done by A than B. There is no violation of the first law, as long as the extra work done by A equals the difference in heats at low temperature. However, we have come up with a combined machine that does useful work by extracting heat from a colder region without changing anything in the hotter region. Were that possible, heat engines would not need any fuel while merrily rolling along, getting all the necessary energy from the surroundings. Such machines are called perpetual motion machines of the second kind and their non-existence is the proof of the Carnot theorem.

Nothing happened for a quarter century after Carnot's publication, perhaps because, as yet, it had nothing to do with practical matters. After 1850, however, the scientific community became interested in thermodynamics and kinetic theory, and research moved into the university halls (the word, "scientist," in fact, did not come into circulation until well after 1833). Clausius, Lord Kelvin, Maxwell, Boltzmann, Gibbs and many other luminaries looked for the scientific theory behind heat engines. Their research did not have a significant impact on heat engine design, but it generated laws that form the basis for investigations of matter in all forms and under all conditions. It is often said, with perfect justification, that science owes more to heat engines than heat engines owe to science.

As scientific interest arose, Lord Kelvin showed that, if the Carnot theorem is valid, there must be an absolute (independent of the nature of substances) temperature scale, which has become known as the Kelvin scale. He also showed that for the Carnot cycle

$$\frac{q_2}{T_2} + \frac{q_1}{T_1} = 0 \tag{6.7}$$

and that Carnot efficiency can also be expressed as

$$\varepsilon = 1 - \frac{T_2}{T_1} \tag{6.8}$$

where T_2 and T_1 are the low and high temperatures respectively. Therefore, the greater the temperature difference the greater the efficiency, but not all heat can be converted unless we can reach absolute zero for the low temperature.

At about the same time, Clausius proved that every reversible cycle will have Carnot efficiency. From that he showed that for any reversible cycle

$$\oint \frac{dq_{rev}}{T} = 0 \tag{6.9}$$

where the integration is over a closed cycle (the initial and the final limits are the same). These scientists, let us keep in mind, were still theorizing about engines that could not be built.

We know that work or heat in a cycle cannot be zero; if it were, there could be no heat engines. Only state functions, which depend on the variables for the state of the system (and not on how that state is reached), have the property shown by the above integral. In 1850 Clausius showed that energy is a state function (first law). Fifteen years later he showed that

$$dS = \frac{dq_{rev}}{T} \tag{6.10}$$

must be a differential form of another state function. He called the state function entropy and gave it the symbol S. Clausius also showed that in an irreversible process

$$dS > \frac{dq}{T} \qquad (6.11)$$

Thus entropy can increase without heat being transferred between the system and the surrounding. This is what happens in an isolated system.

If you don't see the importance of Clausius' discovery of the second state function, don't worry. According to historians, Clausius didn't either. He continued to think of work and heat as primary quantities, energy as adiabatic work (work without heat transfer) and entropy as a measure of unavailable work. This doesn't reflect on his intellectual abilities, which were formidable, but illustrates how difficult it is to get out of thermodynamic cycles once you get into them. I will remind you in Section 7.3 how an excessive emphasis on heat and work (processes) still haunts pedagogy in thermodynamics and makes it difficult to understand important thermodynamic relations.

It was Gibbs (1873) who saw clearly what it meant to have the second state function. He came up with Eq. (6.4), which he called the **fundamental equation of thermodynamics**, by combining the Clausius equation (Eq. (6.10)) and the formula for reversible work.

Gibbs pointed out that Eq. (6.4) is the formula of a surface in USV coordinate space (Fig. 6.3). Every equilibrium point of the system is on this surface. Therefore, we can calculate how properties of the system change from point to point (from state to state) by integrating over a path on this surface. True, the theory has to be done only along unrealistic (in the experimental sense) reversible paths on the surface. But what difference does it make as long as we stick to state functions? The change in state function between any two points must be the same whatever path we use for the calculations. So let the experiment go on from point a to point b under irreversible conditions. If the experimenter waits long enough at point

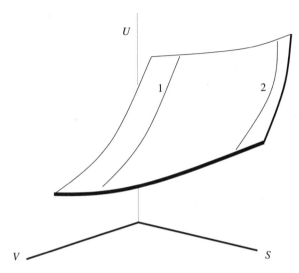

Fig. 6.3. Equilibrium thermodynamic surface in USV coordinate system. This diagram is only for illustration and does not represent any real system. Thermodynamic equations are best understood as relations between slopes on thermodynamic surfaces. Lines 1 and 2 in the diagram are $(\partial U/\partial V)_S$ curves. Notice that $(\partial U/\partial V)_S$ is still a function of S, since the slope for curves 1 and 2 are not the same. Similarly the partial derivative in Eq. (6.18), $(\partial P/\partial T)_V$, is a function of V.

b for a new equilibrium to be established, the results should be the same as those calculated by a theoretician assuming reversibility. One takes the high road and the other, the low road but they meet at the same point. All that is required is that properties of interest be functions of energy and entropy and not of work and heat.

Gibbs was concerned with finding, in his own words, "a general graphical method which can *exhibit at once* all the thermodynamic properties of fluid concerned in *reversible processes*, and serve alike for the demonstration of *general theorems* and the *numerical solution* of particular problems."

I want to show an application of this immediately, so that you do not see thermodynamics as a string of mathematical mysteries. The necessary theory will be developed along the way. You will see that the italicized phrases (my emphasis) in the quotation are connected to the steps in the following example.

6.3. Intermolecular Energy

Attraction or repulsion between molecules determines many properties of chemicals (e.g. viscosity and diffusion). A measure of intermolecular energy, U_{int}, is represented by the partial derivative

$$U_{int} = \left(\frac{\partial U}{\partial V}\right)_T \tag{6.12}$$

If this quantity is negative there is an attraction between molecules; if it is positive, repulsion. As intermolecular distance decreases, attraction turns to repulsion and the sign of the above partial derivative changes. Therefore let us look for a relation between the above partial derivative and experimental data.

Equation (6.4) shows that energy is a function of entropy and volume: $U = U(S, V)$. Hence we have the following expression for the change in energy:

$$dU = \left(\frac{\partial U}{\partial S}\right)_V dS + \left(\frac{\partial U}{\partial V}\right)_S dV \tag{6.13}$$

A comparison of Eqs. (6.4) and (6.13) shows that

$$\left(\frac{\partial U}{\partial S}\right)_V = T; \quad \left(\frac{\partial U}{\partial V}\right)_S = -P \tag{6.14}$$

Since T and P are readily measured, equating them to partial derivatives would appear to be raising the obvious to the obscure. We do not need the partial derivatives to estimate T and P. However, if we visualize energy as a surface in the S−V plane, we see that the partial derivatives are lines on the surface, as illustrated in Fig. 6.3. You will realize, as we go along, that visualizing energy as a surface makes thermodynamic theory both interesting and understandable.

From Eqs. (6.13) and (6.14), we get

$$\left(\frac{\partial U}{\partial V}\right)_T = \left(\frac{\partial U}{\partial S}\right)_V \left(\frac{\partial S}{\partial V}\right)_T + \left(\frac{\partial U}{\partial V}\right)_S$$

$$= T\left(\frac{\partial S}{\partial V}\right)_T - P \tag{6.15}$$

We have not obviously gained anything here since $(\partial S/\partial V)_T$, on the right side of the equation, is an unfamiliar quantity. However, help is on the way. When the rest of the theory is developed, you will see that this obscure partial derivative is related to a partial derivative for a measurable property (increase in pressure with temperature at constant volume):

$$\left(\frac{\partial S}{\partial V}\right)_T = \left(\frac{\partial P}{\partial T}\right)_V \tag{6.16}$$

(This is from Eq. (7.3 M) in the next chapter.)

Hence Eq. (6.15) becomes

$$\left(\frac{\partial U}{\partial V}\right)_T = T\left(\frac{\partial P}{\partial T}\right)_V - P \tag{6.17}$$

Now we see that intermolecular energy can be deduced from just PVT data. The important point is that thermodynamic theory allows us to deduce properties of interest to chemists and material scientists from PVT data. Referring back to the quotation, we have demonstrated a *general theorem* (the equation is valid for all substances) by a *reversible process*.

At this stage we have a choice as to how to proceed. We can (a) measure the quantities on the right side of the equation and evaluate the integral

$$\int dU_T = \int\left[T\left(\frac{\partial P}{\partial T}\right)_V - P\right]dV \tag{6.18}$$

between selected limits or (b) use an *equation of state* to simplify the expression in braces.

For option (a), consider measurement first. The partial derivative $(\partial P/\partial T)_V$ is a function of volume. The subscript only indicates that during measurement, V remains constant. However, you have to take measurements at different values of V (see Fig. 6.3).

Measurements of $(\partial P/\partial T)_V$ have been done on liquids. They involve, however, a great deal of labor since the partial derivative has to be determined at widely different values of P, V and T before Eq. (6.18) can be integrated. So let us consider the alternative, option (b).

The state of a system is determined by the variables P, V and T. Equations of state are mathematical functions that express the relation between these variables. One familiar example is the ideal gas equation of state

$$P = \frac{nRT}{V} \qquad (6.19)$$

where n is the number of moles and R the gas constant. From this equation, and Eq. (6.17), it is readily shown that intermolecular interaction vanishes for the ideal gas. However, it should be noted that equations of state are not derived from thermodynamic laws. They are mathematical constructs that fit the data and give insight into the properties of gases.

An equation of state developed by van der Waals in 1873 has been most useful in understanding behavior of real gases and liquids. The van der Waals equation, with two constants a and b

$$P = \frac{nRT}{V - nb} - \frac{an^2}{V^2} \qquad (6.20)$$

modifies the ideal gas law by incorporating intermolecular forces. The volume available for molecules to move around will be less than the total volume, since molecules themselves occupy space, making the effective volume $V - nb$. Intermolecular attraction reduces the effective pressure exerted by the gas. Interaction between pairs of molecules is proportional to the square of their concentration, an^2/V^2. Hence $(P + an^2/V^2)$, where P is the actual pressure, replaces ideal pressure in Eq. (6.19).

Let us see how the van der Waals equation, in conjunction with Eq. (6.17), allows us to explore intermolecular energy. From Eq. (6.20) we have

$$\left(\frac{\partial P}{\partial T}\right)_V = \frac{nR}{V - nb} \qquad (6.21)$$

and

$$T\left(\frac{\partial P}{\partial T}\right)_V - P = \frac{an^2}{V^2} \qquad (6.22)$$

Hence Eq. (6.18) simplifies to

$$\int dU_T = \int \frac{an^2}{V^2} dV \qquad (6.23)$$

What should be the limits of integration? As $V \to \infty$, intermolecular energy goes to zero; the smallest volume available for a mole of gas is b, and so it is appropriate to take these as the limits.

The result of integration between these limits is

$$\Delta U_{\text{int}} = -\frac{a}{b} \qquad (6.24)$$

Referring back to the quotation, we now have a *numerical solution* since there are methods of determining a and b.

We could also have used the molar volume of the liquid, instead of b, for the lower limit. The main point of this exercise is to show how equations of state are used in studying molecular properties. In their actual application we must take into account the available data and equations of state and approximations suitable for the system under consideration.

Table 6.1 gives the boiling points, T_B, and the intermolecular energies computed from the van der Waals equation for selected substances. Roughly speaking, boiling point is a measure of energy needed to separate molecules into the gas phase, i.e. to overcome intermolecular attraction. Hence the correlation between van der Waals energies and boiling points, as indicated by the data in Table 6.1, is what we would expect.

Table 6.1. This table shows a correlation between boiling point, which is a measure of intermolecular energy, and intermolecular energy calculated from the van der Waals equation.

Compound	a/atm L^2 mol^{-2}	b/L mol^{-1}	T_b/K^1	$-\Delta U_{\text{int}}$/kJ mol^{-1}
Helium	0.0346	0.0238	4.2	0.15
Oxygen	1.36	0.0318	90.2	4.33
Ethane	5.58	0.0651	184.6	8.68
CS$_2$	11.25	0.0726	319.2	15.70
Hexane	24.84	0.1744	341.9	14.43
Perfluorohexane (C$_6$F$_{14}$)	31.03	0.2497	329.2	12.59

Chemists do not always look for the most accurate theory. They often apply theories to a whole series of molecules and use them to predict a pattern in their properties, learning as much from the deviations as from the trends. A good example is the "discovery" of the hydrogen bond in water from boiling point trends in hydrogen compounds of the VI group elements. Water has a far higher boiling point than expected from the trend, indicating specific and relatively strong interaction between water molecules.

The data in Table 6.1 indicate that intermolecular forces are weaker in perfluorohexane than in hexane, even though the constant, a, indicates it should be otherwise. The reason for this becomes clear when we compare the constant b for both. It is the size of the molecules here that keeps them apart and reduces the intermolecular energy, a useful insight we were able to gain from a simple theory.

6.4. Nature of Theory in Thermodynamics

This is an appropriate point to learn about the relation between theory and experiment. The following diagram indicates several connections.

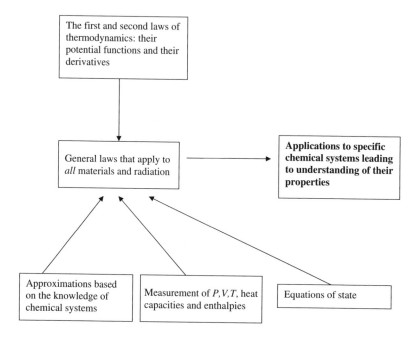

The general laws deduced from thermodynamic theory contain no approximations, but nor do they directly touch the experimental plane. To make that connection, we have to have input from one of the boxes shown at the bottom of the diagram. By remaining detached from the experimental plane, the fundamental theory of thermodynamics has been a sure guide to experimentalists in every area of science and engineering. In areas outside of science, too, it has given guidance; for example, Samuelson developed his theory of economic equilibrium based on Gibbs' formulation of thermodynamics. Think of the fundamental theory as a lighthouse. It tells you what direction to go in, but not the course you have to take.

6.5. What is Actually Measured?

The formal development of thermodynamics depends on abstract concepts and manipulation of partial derivatives. The end results are general equations that contain measurable quantities and apply for all materials. For instance, Eq. (6.17) is valid for all materials and relates intermolecular energy (a property) to the variables of the system. Equation (6.17) is also called the **thermodynamic equation of state** to emphasize its universal nature. As it gives general equations (we could call them laws) and directly connects to experiments, thermodynamics finds application in every area of science and engineering. It is important, then, that we examine the types of data used in the application of thermodynamics to avoid confusion between abstract theoretical quantities and mundane experimental measurements.

6.5.1. *Pressure, volume and temperature*

Data on P, V and T are often listed as coefficient of thermal expansion, α

$$\alpha = \frac{1}{V} \left(\frac{\partial V}{\partial T} \right)_P \tag{6.25}$$

and isothermal compressibility, κ

$$\kappa = -\frac{1}{V}\left(\frac{\partial V}{\partial P}\right)_T \tag{6.26}$$

Listing these quantities instead of P, V, and T makes tables compact. Besides, these partial derivatives often appear in the theoretical formula. Note that the partial derivative in Eq. (6.26) is a negative quantity. This is so that κ can be listed as a positive number in tables. The partial derivative $(\partial P/\partial T)_V$ is not usually tabulated and does not even have a name. However, it can be estimated easily from α and κ. There is a cyclic relation between the partial derivatives of the three variables as shown by the following equation.

$$\left(\frac{\partial P}{\partial T}\right)_V \left(\frac{\partial T}{\partial V}\right)_P \left(\frac{\partial V}{\partial P}\right)_T = -1 \tag{6.27}$$

You may check the validity of this relation starting with the ideal gas equation of state. From Eq. (6.27) and the previous two equations we have

$$\left(\frac{\partial P}{\partial T}\right)_V = -1\left(\frac{\partial V}{\partial T}\right)_P \left(\frac{\partial P}{\partial V}\right)_T = \frac{\alpha}{\kappa} \tag{6.28}$$

6.5.2. *Heat*

If you wrap a wire with resistance around a sample and pass a current through it, you will transfer heat to the sample. If the sample and the heating wire are isolated, heat transferred is simply the product of the power and time (1 joule = 1 watt).

If the substance is in a single phase (solid, liquid or gas), its temperature rises as heat is transferred, so in practice heating is done under quasi-static conditions. Thus we define heat absorbed at constant volume for infinitesimal change in temperature by a quantity called heat capacity at constant volume (C_V)

$$C_V = \frac{\text{Limit}}{\delta T \to 0}\left(\frac{q_V}{\delta T}\right) \tag{6.29}$$

From Eq. (6.4) we see that, if volume does not change, no work is done and heat transferred is exactly equal to energy change. Hence the above equation becomes

$$C_V = \left(\frac{\partial U}{\partial T}\right)_V = T\left(\frac{\partial S}{\partial T}\right)_V \qquad (6.30)$$

The last step follows from Eq. (6.4). We will take the above equation as the definition of C_V since it relates heat absorption to a property of the system.

From Eq. (6.4) and the formula for work it follows that at *constant pressure*

$$q_P = \Delta U + P\Delta V = \Delta(U + PV) \qquad (6.31)$$

Here we are equating a process, heat absorption at fixed pressure, to change in a property, $(U+PV)$, of the system. The quantity $(U+PV)$ is given the symbol H and the name enthalpy.

$$H \equiv U + PV \qquad (6.32)$$

I will further explain the reason for defining the enthalpy function in Section 8.3.

When two phases coexist, transferring heat to the system does not change its temperature. (The temperature of water in equilibrium with steam remains constant as long as both phases are present). In this case we measure the quantity of heat transferred at constant temperature and pressure. Black (1776) gave the descriptive name latent (stored or dormant) heat for this quantity. To avoid the implication that heat is a stored quantity, and to distinguish between a process and a property, it is now referred to as enthalpy change. Thus during phase transition

$$q_P = \Delta H \qquad (6.33)$$

[Heat in transition → Change in property of the system]

In contrast to phase changes, the temperature of a substance in single phase increases when it is heated. Heat capacity at constant pressure

(C_P) is given by the formula

$$C_P = \left(\frac{\partial H}{\partial T}\right)_P \tag{6.34}$$

From Eqs. (6.32) and (6.4), we have

$$dH = dU + PdV + VdP = TdS - PdV + PdV + VdP$$

$$= TdS + VdP \tag{6.35}$$

We must wait until the next chapter to explain the significance of this equation. For now, note that from Eqs. (6.34) and (6.35) we have

$$\boxed{C_P = T\left(\frac{\partial S}{\partial T}\right)_P} \tag{6.36}$$

In summary, here is the list of the quantities measured in thermodynamic studies (in favorable cases they can be calculated from statistical thermodynamic formulas).

1. Pressure, volume, temperature or their partial derivatives.
2. Heat capacities at constant volume and constant pressure.
3. Heat at constant volume (energy change) and at constant pressure (enthalpy change).
4. Energy and enthalpy changes in chemical reactions in a closed vessel called a reaction calorimeter.

Table 6.2 gives heat capacities, enthalpies and entropies for a few substances.

6.5.3. *Heat and radiation*

The standard, and largely very useful, explanation of heat transfer is that it happens through radiation, conduction and convection. After the advent of quantum theory, however, it became clear that all three modes of heat transfer have much in common. Radiation consists of electromagnetic waves as in the case of light while both conduction and convection involve collisions in the broad sense. Collisions, according to quantum theory, are not particles hitting

Table 6.2. Enthalpy, entropy and heat capacity for selected compounds at 298.15 K and 1 bar. Thermochemical data are known for many more significant figures than those indicated in these tables. The numbers here and in Table 6.3 are intended only to give a feeling for the magnitudes involved.

Compound	$\Delta H^\circ/\,\mathrm{kJ\,mol^{-1}}$	$S^\circ/\mathrm{J\,K^{-1}\,mol^{-1}}$	$C_P^\circ/\mathrm{J\,K^{-1}\,mol^{-1}}$
$H_2(g)$	0	115	21
$H_2O(g)$	-242	189	34
$N_2(g)$	0	192	29
$NO(g)$	90	211	30
$NH_3(g)$	-46	192	35

Superscript $^\circ$ indicates that the values are at 298.15 K and 1 bar. Entropy is calculated from heat capacity data as indicated in Fig. 6.4. Enthalpy and energy can only be stated as differences from standard states. Elements in their stable form are assigned zero enthalpy (H_2 and N_2 above). Enthalpy of a compound is calculated from heat capacity data, taking elements as the base line.

each other like billiard balls. When two atoms or molecules (which have oscillating positive and negative charges) go past each other, the electric fields they generate are a superposition of electromagnetic waves. The interaction is mediated by absorption and emission of these waves during the transit. Collision then, to use the colorful language of physicists, is the exchange of "virtual photons."

According to this view, heat absorption is very much like a spectroscopic transition. In the same way they absorb light, molecules absorb heat when the frequencies in radiation match energy differences in the molecules. Thus molecules with widely separated energy levels have lower heat capacity. Molecules with densely packed energy levels, on the other hand, absorb more frequencies from the radiation and have higher heat capacity.

In a seminal paper in 1907, Einstein used the connection between heat capacity and energy states to lay the foundation for quantum theory. The idea is both simple and profound. The available data indicated that the heat capacity of atomic crystals vanishes as temperature goes to zero. This could only happen if energy is quantized and radiation at very low temperature does not have the frequencies to cause a transition between quantum states. If the energy of the crystal is a continuum, as classical theory assumes, the crystal should be able to absorb whatever frequencies there are in the radiation and thus even at absolute zero its heat capacity would not be zero, contrary to experimental results.

6.6. What is Deduced?

6.6.1. *Entropy and enthalpy*

Clausius' equation (Eqs. (6.9) and (6.10)) shows the connection between heat and entropy. From that equation it follows that entropy change at constant volume is given by

$$\int dS(\text{single phase}) = \int \frac{C_V}{T} dT \qquad (6.37)$$

and at constant pressure by

$$\int dS(\text{single phase}) = \int \frac{C_P}{T} dT \qquad (6.38)$$

Heat absorbed at constant temperature and volume equals energy change. Hence entropy change at constant temperature and volume is given by

$$\Delta S_{V,T}(\text{phase change}) = \frac{\Delta U_V}{T} \qquad (6.39)$$

Entropy change at constant pressure and temperature is given by

$$\Delta S_{P,T}(\text{phase change}) = \frac{\Delta H}{T} = \frac{q_P}{T} \qquad (6.40)$$

Enthalpy and entropy changes during phase transitions for a few substances are given in Table 6.3.

Table 6.3.   Enthalpy and entropy changes during phase transitions.

Compound	$\Delta H/\text{kJ mol}^{-1}$	$\Delta S/\text{J K}^{-1}\text{mol}^{-1}$
$H_2O\ (l) \rightarrow H_2O(g)$	41	118
$H_2O\ (s) \rightarrow H_2O(l)$	6	22
$C_6H_6(l) \rightarrow C_6H_6(g)$	31	88
$C_6H_6(s) \rightarrow C_6H_6(l)$	11	39

6.6.2. *Absolute entropies*

If we assume that entropy is zero at 0 K, we can compute the absolute entropy for a substance. According to the third law of thermodynamics entropy at 0 K is either zero or a very small number as we will see in Chapter 9.

Figure 6.4 illustrates how entropy of O_2 at 350 K was determined. We start with the condition $S(0) = 0$. Heat capacity data are not available between 0 and 14 K. In this region we depend on a theory of heat capacity developed by Debye, which shows that heat capacity varies as a third power of T. Beyond 14 K, we use Eq. (6.38) for a pure phase and Eq. (6.40) for phase change.

6.6.3. *Work*

The second term on the right side of Eq. (6.4) does not correspond to the actual work in engineering application. This is because P in

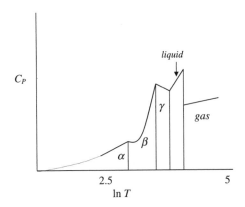

Fig. 6.4. Heat capacity of oxygen from near 0 K to 350 K. The three solid forms of oxygen are known as α, β, and γ forms. For entropy change with temperature in a pure phase, we use the formula

$$\Delta S = \int C_P \ln T$$

The formula for entropy change during a phase transition is

$$\Delta S = \frac{\Delta H (\text{phase change})}{T} = \frac{q_P}{T}$$

Enthalpy change within a phase is obtained from the formula $\Delta H = \int C_P dT$. For phase transitions it is measured as q_P.

Eq. (6.4) is the pressure of the substance in the system, and not the external pressure, P_{ext}. Work done by the system on the surroundings is given by

$$\text{work} = -\int P_{ext} dV \qquad (6.41)$$

If $P_{ext} = 0$, we get no work even though the system expands. Work increases as external pressure approaches the system pressure. Hence the work term in Eq. (6.4) gives the maximum possible work. It is also called reversible work since it corresponds to a (hypothetical) process in which the system and surroundings continue to be in equilibrium while the system is expanding or contracting.

Reversible work is always computed from PVT data or from an equation of state. Actual work is estimated from an indicator diagram, a plot of P_{ext} against volume of the *system*, as shown in Fig. 6.1.

This chapter started with the introduction of the fundamental equation of thermodynamics. The rest of thermodynamic theory follows from this equation. We have considered just one example of how this equation is used. It illustrates how thermodynamic theory "comes down" to the experimental plane. The description of which quantities are measured and which are deduced *must* be clearly understood to avoid confusion.

 Related material on the disc.

With CD-ROM

Chapter 7

Thermodynamic Potentials

The fundamental equation,

$$dU = TdS - PdV \qquad (7.1)$$

represents an energy surface with S and V as the coordinates. Are there other surfaces in different coordinate systems that we might find useful in the study of thermodynamic properties? The answer is yes as we will see in this chapter.

7.1. Important Functions

7.1.1. *Helmholtz energy*

The curve in Fig. 7.1 is a line on the energy surface in Fig. 6.3. It shows energy as a function of entropy at a particular volume. The curve is concave since

$$\left(\frac{\partial U}{\partial S} \right)_V = T > 0$$

Consider an *isolated system*, not at equilibrium, represented by point X. Since the system is isolated, it cannot interact mechanically or thermally with the surroundings and can reach equilibrium only by an increase in entropy. This is shown by the horizontal arrow from X to S_1 in Fig. 7.1. (All equilibrium states are on the curve

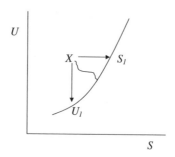

Fig. 7.1.   Transition from a non-equilibrium state to an equilibrium state in an energy–entropy plane. See text for discussion.

and the non-equilibrium states, to the left of the curve. There can be no states to the right of the curve, since entropy cannot decrease in an isolated system.)

Now, if we keep the entropy constant at X (volume is also constant), the system can only reach an equilibrium state by a decrease in energy, as shown by the arrow X to U_1. You might think that energy cannot change if both S and V are fixed. While that is certainly true if the system is at equilibrium, i.e. at a point on the U–S curve, point X is not on the equilibrium surface. It has *at least* three degrees of freedom: U, S and V. It may also have others, since density and temperature could vary from region to region. Hence both entropy and volume may remain constant while energy decreases.

What happens if we don't control either U or S but keep volume constant? The system reaches an equilibrium point between U_1 and S_1, as shown by the wiggly line. Thus the function that is minimized *with respect to non-equilibrium states* is neither U nor S, but a combination of both. From dimensional analysis we see that this function has to be

$$U - TS = A \qquad (7.3)$$

since the product of T and S has dimensions of energy. A is called **Helmholtz energy** or **Helmholtz potential**. (Note: The equation references that follow are not consecutive; the references match those in Table 7.1.)

Table 7.1. Thermodynamic potentials and their partial derivatives.

Number		Equation
Eq. (7.1)	The fundamental equation of thermodynamics	$dU = TdS - PdV$
Thermodynamic Potentials		
Eq. (7.2)	Enthalpy	$H = U + PV$
Eq. (7.3)	Helmholtz energy	$A = U - TS$
Eq. (7.4)	Gibbs energy	$G = H - TS$
Differential forms		
Eq. (7.2 D)	Enthalpy	$dH = TdS + VdP$
Eq. (7.3 D)	Helmholtz energy	$dA = -SdT - PdV$
Eq. (7.4 D)	Gibbs energy	$dG = -SdT + VdP$
Principal *Slopes*		
Eq. (7.1 S)	Energy	$\left(\dfrac{\partial U}{\partial S}\right)_V = T; \quad \left(\dfrac{\partial U}{\partial V}\right)_S = -P$
Eq. (7.2 S)	Enthalpy	$\left(\dfrac{\partial H}{\partial S}\right)_P = T; \quad \left(\dfrac{\partial H}{\partial P}\right)_S = V$
Eq. (7.3 S)	Helmholtz energy	$\left(\dfrac{\partial A}{\partial T}\right)_V = -S; \quad \left(\dfrac{\partial A}{\partial V}\right)_T = -P$
Eq. (7.4 S)	Gibbs energy	$\left(\dfrac{\partial G}{\partial T}\right)_P = -S; \quad \left(\dfrac{\partial G}{\partial P}\right)_T = V$
Maxwell's Relations		
Eq. (7.1 M)	From U	$\left(\dfrac{\partial T}{\partial V}\right)_S = -\left(\dfrac{\partial P}{\partial S}\right)_V$
Eq. (7.2 M)	H	$\left(\dfrac{\partial T}{\partial P}\right)_S = \left(\dfrac{\partial V}{\partial S}\right)_P$
Eq. (7.3 M)	A	$\left(\dfrac{\partial S}{\partial V}\right)_T = \left(\dfrac{\partial P}{\partial T}\right)_V$
Eq. (7.4 M)	G	$\left(\dfrac{\partial S}{\partial P}\right)_T = -\left(\dfrac{\partial V}{\partial T}\right)_P$

In equation numbering in the text D is used for the differential form, S, for the first slope and M, for Maxwell's relation.

In the differential form

$$dA = dU - TdS - SdT = TdS - PdV - TdS - SdT$$
$$= -SdT - PdV \tag{7.3 D}$$

Note that the Helmholtz energy is a function of T and V.

It directly follows from the above equation that entropy and pressure are

$$\left(\frac{\partial A}{\partial T}\right)_V = -S \quad \text{and} \quad \left(\frac{\partial A}{\partial V}\right)_T = -P \tag{7.3 S}$$

and that the partial derivatives of S and P must have the following property:

$$\left(\frac{\partial S}{\partial V}\right)_T = \left(\frac{\partial P}{\partial T}\right)_V \tag{7.3 M}$$

Equation (7.3 M) is one of the four equations that are usually referred to as Maxwell's relations.

If we write Eq. (7.3 M) in an expanded form,

$$\left(\frac{\partial}{\partial V}\left(\frac{\partial A}{\partial T}\right)_V\right)_T = \left(\frac{\partial}{\partial T}\left(\frac{\partial A}{\partial V}\right)_T\right)_V$$

we see that the order of differentiation does not matter. This is the essential idea behind Maxwell's relations.

There are two other functions to look at before we can connect the theory of these equations to experiment.

7.1.2. Enthalpy

The curve in Fig. 7.2 is a line on the energy surface in Fig. 6.3. X again represents the non-equilibrium point considered above. Note

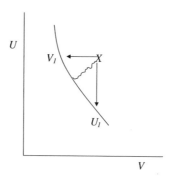

Fig. 7.2. Transition from a non-equilibrium state to an equilibrium state in energy–volume plane. See text for discussion.

that the positive S-axis is behind the page. Hence X has to be above the curve. How does it reach equilibrium if V is fixed? The line X to U_1 in Fig. 7.2 shows that energy decreases in this transition. If we keep U and S constant the system can reach equilibrium only by decrease in volume, as shown by the line from X to V_1. Thus both energy and volume change in the same direction when the system moves toward equilibrium. If, however, we allow both V and U to vary at constant entropy, the system will reach equilibrium when the function

$$H = U + PV \qquad (7.2)$$

(**enthalpy** Eq. (6.32)) reaches a point on the equilibrium curve, as shown by the wiggly line in the figure. Enthalpy in the differential form is given by

$$dH = TdS + VdP \qquad (7.2\,\mathrm{D})$$

(see also Eq. (6.35)).

7.1.3. *Gibbs energy*

From the above two examples it should be evident that if U, V and S were to vary, the system would reach equilibrium when the following function, called the Gibbs energy, were a minimum with respect to all other non-equilibrium states

$$G = U + PV - TS = H - TS \qquad (7.4)$$

Figure 7.3 gives further explanation.

In the differential form, Eq. (7.4) Gibbs energy is

$$dG = -SdT + VdP \qquad (7.4\,\mathrm{D})$$

We now have three functions besides the energy function. Table 7.1 gives the mathematical properties of these functions.

I have introduced four functions in a quick succession. What do they mean? I will give two applications before answering this question.

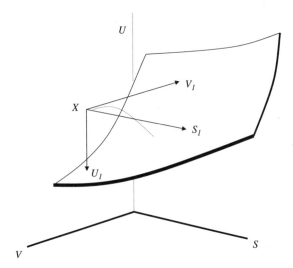

Fig. 7.3.   Transition from a non-equilibrium state to an equilibrium state on the energy–entropy–volume surface.

7.2. Applications

We have already found use for the Maxwell relation (Eq. (7.3 M)) in the previous chapter (Eq. (6.16)) in converting a partial derivative of entropy to a partial derivative of pressure. Let us consider other examples.

7.2.1. *Entropy change*

Is there a general formula for change of entropy with volume at a fixed temperature? The answer drops out of Maxwell's relation (Eq. (7.3 M)).

$$\int dS_T = \int \left(\frac{\partial P}{\partial T}\right)_V dV \tag{7.5}$$

This is a formula unadulterated by approximations. If you come across a strange form of matter, hitherto unsuspected, you can still use this equation. The procedure is to measure the partial derivative at different volumes and integrate the resulting expression, but if you happened to have an equation of state, you could save yourself the labor. For an ideal gas the above equation simplifies to

$$\Delta S_T = nR \ln \frac{V_2}{V_1} \tag{7.6}$$

The subscripts indicate the limits of integration.

For a van der Waals gas

$$\Delta S_T = nR \ln \frac{(V_2 - nb)}{(V_1 - nb)} \tag{7.7}$$

We get a better feeling for entropy changes by expanding the above expression. Let us make $n = 1$ for convenience.

$$\Delta S_T = \left[R \ln V_2 + R \ln \left(1 - \frac{b}{V_2} \right) \right] - \left[R \ln V_1 + R \ln \left(1 - \frac{b}{V_1} \right) \right]$$

$$\approx \left[R \ln V_2 - \left(\frac{b}{V_2} \right) \right] - \left[R \ln V_1 - \left(\frac{b}{V_1} \right) \right] \tag{7.8}$$

Entropy at both volumes is less than the ideal entropy since the available volume for the molecules is reduced. If $V_1 < V_2$, there is greater reduction of entropy at smaller volume than at a larger volume, because excluded volume becomes more significant as volume decreases. Hence entropy change between V_2 and V_1 in the non-ideal gases may be larger than in ideal gases. Abrupt changes in the optical properties of colloidal solutions have been observed at certain volume fractions. This is attributed partly to change in the available volume due to aggregation and thus entropy.

Equations (7.6), (7.7) and (7.8) are valid only for gases. If we want to estimate entropy change in liquids or solids, we have to start with Eq. (7.5). When the partial derivative in the integral is deduced from the coefficient of thermal expansion, α (Eq. (6.25)), and isothermal compressibility, κ (Eq. (6.26)) we get

$$\int dS_T = \int \frac{\alpha}{\kappa} dV \tag{7.9}$$

From the α and κ values listed in standard tables, we can estimate that entropy changes in water and benzene at 298 K are $0.077 \, \text{J K}^{-1} \text{mol}^{-1}$ and $1.2 \, \text{J K}^{-1} \text{mol}^{-1}$ when their molar volumes increase by 1% and pressure decreases correspondingly (keep in mind that a partial derivative is still a function of the variable indicated as constant in the subscript). Hydrogen bonds in liquid water restrict the freedom of molecules to move around, compared to molecules in benzene. The entropy changes are indicative of a more "ordered" arrangement of molecules in water.

7.2.2. Difference in heat capacities

It is often more convenient to measure heat capacity at constant pressure. For interpretation of the results on a molecular level, however, it is useful to have heat capacity at constant volume, since the latter is directly related to energy. The theory developed above is useful in computing one heat capacity from the other, so I will give here every step in the derivation. From Eqs. (6.30) and (6.36) we have

$$\frac{C_P - C_V}{T} = \left(\frac{\partial S}{\partial T}\right)_V - \left(\frac{\partial S}{\partial T}\right)_P \tag{7.10}$$

There is no Maxwell's relation to convert the above partial differentials. For that we need S as a function of V and T, or P and T. Let us try V and T as the variables. Then

$$dS = \left(\frac{\partial S}{\partial T}\right)_V dT + \left(\frac{\partial S}{\partial V}\right)_T dV \tag{7.11}$$

From this equation we have (by further differentiation with respect to T)

$$\left(\frac{\partial S}{\partial T}\right)_P = \left(\frac{\partial S}{\partial T}\right)_V + \left(\frac{\partial S}{\partial V}\right)_T \left(\frac{\partial V}{\partial T}\right)_P \tag{7.12}$$

Hence

$$\left(\frac{\partial S}{\partial T}\right)_P - \left(\frac{\partial S}{\partial T}\right)_V = \left(\frac{\partial S}{\partial V}\right)_T \left(\frac{\partial V}{\partial T}\right)_P \tag{7.12 b}$$

Now the first partial derivative on the right side in the above equation can be replaced by Maxwell's relation, Eq. (7.3 M) $[(\partial S/\partial V)_T =$

$(\partial P/\partial T)_V$] and the left side with the heat capacity difference from Eq. (7.10). Hence we have

$$C_P - C_V = T \left(\frac{\partial P}{\partial T}\right)_V \left(\frac{\partial V}{\partial T}\right)_P \qquad (7.13)$$

According to Eq. (6.28), the first partial derivate on the right side is (α/κ) where α is the coefficient of thermal expansion and κ, the compressibility. The second partial derivative in the above equation is $V\alpha$. Hence we have

$$C_V = C_P - \frac{TV\alpha^2}{\kappa} \qquad (7.14)$$

For an ideal gas, the right-hand side reduces to the gas constant R. For other substances we have to find the data for the partial derivatives on the right-hand side. Take solid copper as an example: at 300 K, $C_P = 24.5$ and $C_V = 23.8 \, \mathrm{J\,K^{-1}mol^{-1}}$; at 1200 K, $C_P = 30.2$ and $C_V = 26.0 \, \mathrm{J\,K^{-1}mol^{-1}}$.

To see why we need Eq. (7.14), think of the way experiments have to be conducted. To measure C_P we can heat the sample at atmospheric pressure. To measure C_V, on the other hand, we have to keep the volume constant, which can only be done by gradually compressing the sample to compensate for thermal expansion; not an easy experiment. It makes more sense to deduce C_V from C_P.

7.3. What Do They Mean?

We have gone through a few mathematical exercises deriving useful relations. These, and many other applications of thermodynamic theory, depend on the definition of enthalpy, Helmholtz and Gibbs energies and their mathematical properties. What do these functions actually represent? How could we describe them in a plain language?

Mathematics connects to science on several levels. On one level, mathematics is used just to represent data. Equations of state are examples of this. They are deduced to fit some already established data and then extended to a wider domain. On another level, quite removed from the first, some mathematical expressions stay suspended above the experimental plane. They connect to the experiments so indirectly that they cannot be identified with a particular experiment. Consider, for instance, $\sqrt{-1}$. Does it have a direct link to anything you see in the lab? Yet it is used in the mathematical plane, to connect one equation to another. The phase of the wave function is often an imaginary quantity. The wave function itself is related to probability density, a real number, which can be measured by scattering techniques. Thus the connection between $\sqrt{-1}$ and experiment, if we want to call it a connection in the first place, is removed by at least one step.

Every mathematical expression that appears in physical sciences is ultimately connected to an experimental quantity and often, to more than one. However, the main purpose of some mathematical entities is to bring the powerful tools of mathematics into scientific theory.

Returning to thermodynamic potentials, I do not deny that they are connected to experiments, but they are not introduced for that purpose. Their main role is to make the theory general and universal. I believe the only honest thing we can say about these potentials is that we know both entropy and energy control all natural processes, so by making potentials from a combination of these quantities, following mathematical rules, we are able to develop a powerful theory that connects one set of experimental results to another.

Similarly, some elementary textbooks define enthalpy as heat absorbed at constant pressure. While this is not wrong, it is quite misleading. If enthalpy is just q_P, why give it another name? Besides, enthalpy is also equal to $\int V dP$ if $dS = 0$. How do we interpret this relation?

Helmholtz energy is often said to be a representation of isothermal work. In fact the symbol A comes from the German word for work, *Arbeit*, but it is not *just* work! The word was coined in the days when scientists were concentrating on processes rather than on state functions. The ghost of the past still lingers. The only explanation that can be defended is that Helmholtz energy is a function that is a minimum at equilibrium (with respect to all non-equilibrium states) when V and T are held constant.

Even some higher-level books fall into the same trap of identifying G with work. After some juggling they state that G is work beyond expansion work. How could that be? Where are the variables in the definition of G that represent that extra work? Certainly we can add the terms that include electrical and magnetic fields, but we can add those terms to U, H or A as well. If G represents *non-expansion* work at constant T and P, H also represents work at constant S and P. Therefore let us just say that analogous to Helmholz energy, Gibbs energy is a function that is a minimum (with respect to all non-equilibrium states) when P and T are held constant.

I believe that thermodynamics is made more difficult by these misleading and meaningless statements and by not properly explaining the mathematical structure of the theory. The importance of these thermodynamic potentials is that they can be manipulated mathematically to give relations between experimental quantities, on one hand, and energy and entropy on the other. From the latter we find explanations for molecular behavior. If you need to visualize something concrete when working with thermodynamic potentials, think of them as mathematical surfaces whose slopes represent thermodynamic properties.

The primary quantities in thermodynamic theory are energy, enthalpy, Helmholtz and Gibbs energies. These are also called thermodynamic potentials. They depend on the variables pressure, temperature and volume. In the theory we have examined, entropy also takes the role of a variable. The power of the theory is its ability to relate many properties of materials to these potentials and derive, through Maxwell's equations, general laws applicable to all systems.

 Related material on the disc.

With CD-ROM

Chapter 8

Chemical Potential

How does the yield of ammonia change with temperature in the Haber process? What determines the solubility of a chemical and its variation with temperature and pressure? The theory that we have covered so far cannot answer these questions, since we have not yet taken chemical composition into account.

Gibbs, in his monumental 1875 paper, extended thermodynamic theory to include chemical composition. This was a major development in science, since most materials are mixtures of substances. He showed that a thermodynamic function, which he called simply "potential," governs the chemical composition. We now call it **chemical potential**.

Gibbs wrote, "One of the principal objects of theoretical research in any department of knowledge is to find the point of view from which the subject appears in its greatest simplicity." That is my aim in this chapter. Section 8.1 develops this "point of view" from which all chemical and phase equilibria appear in their "greatest simplicity."

8.1. Chemical Potential and Equilibrium

We will start with the following equation:

$$G = \sum n_i \mu_i \qquad (8.1)$$

where the subscript denotes a chemical species in the vessel. The equation simply claims that the total Gibbs energy can be partitioned into a sum of Gibbs energies of the components (called chemical potentials and denoted by the symbol μ), as if each component had "private" Gibbs energy. Are we justified in assuming that there are no cross terms involving two different chemicals? The answer is yes, as we will see.

Often difficulties in understanding chemical thermodynamics arise not from the mathematical background but from a misunderstanding of the subtle way in which theory relates to experiment. Let us consider the equilibrium in a two component (A and B) system to examine the connection between theory and experiment:

$$A \leftrightarrow B$$

A and B could be two different phases (e.g. ice and water) or two different chemicals in the same phase. It makes no difference to theory. It makes a huge difference, however, whether it is an equilibrium or a reversible process that we are analyzing. Figure 8.1 shows a line for both equilibrium and reversible changes. At equilibrium point E,

$$dG = 0 \qquad (8.2a)$$

This equation has more than one meaning. Firstly, it says G is a minimum *with respect to all non-equilibrium states under the given environmental conditions*. Since establishing conditions for thermodynamic equilibrium starts with non-equilibrium states, we are often stuck with the wrong idea that G is a minimum on the equilibrium surface. It is not, and we should try to avoid thinking about non-equilibrium states altogether at this stage.

At the equilibrium point E', again $dG = 0$, since environmental variables (usually pressure and temperature) are fixed. The equation now *does not* say anything about a minimum. Thermodynamic equilibrium, as we have noted in Chapter 5, is dynamic. Nothing may

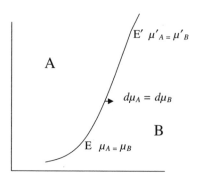

Fig. 8.1. Representation of equilibrium and reversible change in a two-component or two-phase system. Since $dG = 0$ and the number of molecules exchanging roles must be the same ($dn_A = dn_B$), chemical potentials of A and B must be equal for every equilibrium state. Hence during the transition from one equilibrium state to another $d\mu_A = d\mu_B$.

be happening on a macroscopic level but on a microscopic level it is ceaseless motion. Hence, from Eq. (8.1) we get for a two component system,

$$dG = 0 \tag{8.2b}$$

[No change on macroscopic level]

$$(\mu_A dn_A + \mu_B dn_B) + (n_A d\mu_A + n_B d\mu_B) = 0 \tag{8.3}$$

[Change on a molecular level consistent with Eq. (8.2b)]

Since equilibrium is dynamic, at equilibrium A and B molecules exchange continuously. Thus the term in the first brackets corresponds to dynamic equilibrium. The second term is called the Gibbs–Duhem equation, which is used to derive the properties of one component from the properties of the other. These two terms have to vanish independently. So for a dynamic equilibrium

$$\mu_A dn_A + \mu_B dn_B = 0 \tag{8.4}$$

Since the number of molecules of A changing into B must be equal to the number of molecules of B changing into A

$$dn_A = -dn_B$$

and

$$\mu_A = \mu_B \qquad (8.5)$$

[Condition for equilibrium between components]

The above reasoning must also hold for equilibrium point E' in Fig. 8.1. Hence

$$\mu_{A'} = \mu_{B'} \qquad (8.6)$$

From the above two equations it follows that

$$d\mu_A = d\mu_B \qquad (8.7)$$

for reversible change between two equilibrium points.

Our interest lies in finding out how equilibrium changes with pressure, temperature and other variables. The main point is that, if we know an expression for how chemical potential of each component changes $(d\mu)$, we can derive the formulas for equilibrium from Eq. (8.7). It emerges that working with (μ/T), instead of μ alone has some advantages. So Eq. (8.7) is modified to give

$$d\left(\frac{\mu_A}{T}\right) = d\left(\frac{\mu_B}{T}\right) \qquad (8.8)$$

Thus far we have considered only a two component system. What about a reaction like the following?

$$A + 3B \leftrightarrow 2C$$

For this chemical equilibrium, Eq. (8.4) takes the form

$$\mu_A dn_A + \mu_B dn_B + \mu_C dn_C = 0$$

Since $dn_B = 3dn_A$ and $dn_C = -2dn_A$, the condition for equilibrium is

$$\mu_A + 3\mu_B = 2\mu_C \qquad (8.9)$$

For reversible change, it is

$$d\mu_A + 3d\mu_B = 2d\mu_C \qquad (8.10a)$$

and

$$\frac{d\mu_A}{T} + 3\frac{d\mu_B}{T} = 2\frac{d\mu_C}{T} \qquad (8.10b)$$

Let us recap what all this means.

(1) According to Eq. (8.9), if there is an equilibrium between two or many species, the weighted sum of chemical potentials of everything (i.e. chemical species, different phases, ions) on the left-hand side of the equality sign must be equal to the weighted sum of chemical potentials of everything on the right. Let me remind you we are talking about dynamic equilibrium. You see no change in macroscopic variables such as pressure and concentrations, but on the microscopic level, molecules are moving into each other's space and stealing each other's identity continuously.

(2) According to Eqs. (8.10a) and (8.10b), we can find out how equilibrium changes by following changes in chemical potentials.

It is easy to overlook the importance of these statements. Thermal equilibrium requires that temperature be uniform throughout the system. Mechanical equilibrium requires that pressure be uniform throughout the system. Now that we know chemical potentials must be uniform throughout the system for either chemical or phase equilibrium, Eqs. (8.9), (8.10a) and (8.10b) allow us to predict the position of equilibrium and how it varies. The actual calculations may not be elegant but we have a principle that will guide us as we study chemical systems, which, for the most part, is the strength of thermodynamics.

Now, for the important question. How do we find out expressions for chemical potential and the way in which it varies? The derivation is not difficult and is given in the next section, but if you want to go ahead with the result, the following is the expression for variation of $\frac{\mu}{T}$ for component A in an *ideal solution*:

$$d\left(\frac{\mu_A}{T}\right) = -\frac{h_A}{T^2}dT + \frac{v_A}{T}dP + Rd\ln X_A \qquad (8.11)$$

In this equation h stands for molar enthalpy, v for molar volume and X for mole fraction. The subscripts indicate the component. Let us use lowercase letters for molar quantities and avoid using subscripts or superscripts. All the quantities on the right-side can be measured or estimated from experimental data. Enthalpy is determined from heat capacity data or by direct measurement of heat transfer at constant pressure.

8.2. Derivation of Chemical Potential Equation

Before we can find an equation for chemical potential, we have to derive one which shows the relation between Gibbs energy, G, and another function, (G/T). There is no particular name for this important function, (G/T), which has to be given independent status from G. The relation between the two is obtained by implicit differentiation.

$$d\left(\frac{G}{T}\right) = \frac{dG}{T} - \frac{G}{T^2}dT \qquad (8.12)$$

Now, if we substitute Eq. (7.4 D) $(dG = -SdT + VdP)$ for the first term on the right and Eq. (7.4) $(G = H - TS)$ for the second term and simplify, we get

$$d\left(\frac{G}{T}\right) = -\frac{H}{T^2}dT + \frac{V}{T}dP \qquad (8.13)$$

which is the **Gibbs–Helmholtz** equation. This is an equation that applies to the whole system but contains no information on individual chemicals. The only method we know of introducing chemical composition starts with ideal solutions. Non-ideality can be introduced later without revising the theory.

For an ideal solution, both enthalpy and volume are additive. The components behave as if they have private enthalpies and volumes. The only additional term we get is entropy of mixing, which we first encountered in Chapter 5 (Eqs. (5.7) and (5.8)). Hence for a two component system, we have

$$d\left(\frac{G}{T}\right) = n_A d\left[\left(\frac{h_A}{T^2}\right) + \frac{v_A}{T} + R\ln X_A\right]$$

$$+ n_B d\left[\left(\frac{h_B}{T^2}\right) + \frac{v_B}{T} + R\ln X_B\right]$$

$$= n_A\left(\frac{d\mu_A}{T}\right) + n_B\left(\frac{d\mu_B}{T}\right) \qquad (8.14)$$

Thus we have the expression for $\frac{d\mu}{T}$ shown in Eq. (8.11).

My aim is to show, in a few pages, how Eqs. (8.8) (or 8.10b) and (8.11) are the basis for understanding *all chemical and phase equilibria*. Hence I will initially consider only ideal systems, express concentrations only as mole fractions and use the ideal gas law for the vapor phase. What you miss in details will be more than compensated by what you gain by seeing the panorama of chemistry from a single viewpoint.

8.3. Two Phases, One Component

The chemical equation is

$$A(\alpha) \leftrightarrow A(\beta)$$

where α and β denote the different phase of a chemical species A. Evaporation of liquids, sublimation of solids, melting (or fusion) of solids and polymorphism (different solid forms of the same substance) are all examples of phase equilibrium in one-component systems. Since mole fraction of A in each phase is unity, it is not a variable. Thus Eq. (8.8) combined with Eq. (8.11) takes the form

$$-\frac{h_{A,\alpha}}{T^2}dT + \frac{v_{A,\alpha}}{T}dP = -\frac{h_{A,\beta}}{T^2}dT + \frac{v_{A,\beta}}{T}dP \qquad (8.15)$$

By rearranging this equation we get

$$\frac{dP}{dT} = \frac{\Delta h}{T\Delta v} \qquad (8.16)$$

In this equation Δv is the difference in molar volumes of the two phases. The convention is to take the difference between the high-temperature form and the low-temperature form. In the numerator, Δh stands for the enthalpy change (high-temperature form — low-temperature form) in a phase transition. It is easy to forget that this is nothing more than heat absorbed by a mole of substance during phase transition. Both quantities on the right-hand side of the equation are experimentally measurable quantities, so we have a powerful method of predicting phase equilibrium at widely different pressures and temperatures.

Equation (8.16) is known as the Clapeyron equation, which was derived in 1830 from entirely different premises. It can be further simplified for sublimation and evaporation.

The volume of a gas (vapor) is much larger than the volume of a condensed phase. Hence we can approximate

$$\Delta v = v(\text{vapor}) - v(\text{condensed phase}) \approx v(\text{vapor}) = \frac{RT}{P} \quad (8.17)$$

The ideal gas law is used in the last step and works quite well, if the vapor pressure is low. Substitution of Eq. (8.17) into Eq. (8.16) leads to the differential equation

$$d \ln P = \frac{\Delta h}{RT^2} dT \quad (8.18\text{a})$$

$$\ln \frac{P_2}{P_1} = \frac{\Delta h}{R} \left(\frac{1}{T_1} - \frac{1}{T_2} \right) \quad (8.18\text{b})$$

This equation also works for sublimation equilibrium if enthalpy of vaporization is replaced by enthalpy of sublimation. Equation (8.18) is a form of the Clausius–Clapeyron equation.

Let us estimate the boiling point of water at 50 atm pressure. From Eq. (8.18b) we have

$$\ln \frac{50}{1} = \frac{\Delta h}{R} \left(\frac{1}{373} - \frac{1}{T} \right)$$

Enthalpy of vaporization of water is $40.7 \, \text{kJ} \, \text{mol}^{-1}$. Hence from the above equation you get $T = 531 \, \text{K}$ as the boiling point of water at 50 atm, the pressure at the depth of 1500 feet in the ocean.

Hydrothermal vents (hot water geysers) were discovered at depths of 2100 m in mid-ocean ridges. The temperature of the jetting water was estimated to be around 670 K. Since pressure was high (weight of ocean water) water does not boil. Equation (8.18b) cannot be used when there is a large change in temperature, since enthalpy depends on temperature. In this case we start with Eq. (8.18a), substitute temperature-dependent enthalpy function (which is obtained from heat capacity data) before proceeding with integration.

8.4. Two Phases, Two Components

Symbolically we can write this equilibrium as

$$A \text{ and } B(\alpha) \leftrightarrow A(\beta)$$

Examples are vapor pressure of a liquid A when a non-volatile solute B is added to A (Raoult's law), solubility of a gas A in a liquid B (Henry's law), and solubility of a solid A in a liquid solvent B. In all these cases B does not become directly involved in the equilibrium, even though it influences the behavior of A.

8.4.1. *Raoult's law*

In the Raoult's law example, the solvent (we will denote it by subscript 1 to conform to the standard notation) is in the liquid phase. Its mole fraction varies as the solute (subscript 2) is added. The vapor phase consists *only* of solvent vapor, and so its mole fraction is not a variable but pressure is. Thus we have X as the variable in the liquid phase and P as the variable in the vapor phase. Hence Eq. (8.8) combined with Eq. (8.11), with some rearranging gives

$$\frac{v}{RT}dP_1 = d\ln X_1 \qquad (8.19)$$

If the solvent has low vapor pressure, we can use the ideal gas law, $\frac{v}{RT} = \frac{1}{P}$. When $X_1 = 1$ (pure solvent), the vapor pressure is that of the pure solvent, P_1^o. Integration between this limit and some other value of X_1

$$\int_{P^o}^{P_1} d\ln P_1' = \int_1^{X_1} d\ln X_1' \qquad (8.20)$$

gives

$$P_1 = X_1 P_1^o \qquad (8.21)$$

where P_1^o is the vapor pressure of the pure liquid.

As you add a non-volatile solute to a solvent, the mole fraction of solvent and its vapor pressure decrease. For a two component system, $X_1 + X_2 = 1$. This allows us to rearrange Eq. (8.13) into

$$\frac{P_1^o - P_1}{P_1^o} = X_2 \qquad (8.22)$$

In this form Raoult's law is used to determine molar masses of non-volatile substances. The procedure is to dissolve a known weight of a non-volatile solute in a volatile solvent and determine the mole fraction of the solute by the above formula. From the mole fraction and the weight of the sample it is a simple matter to calculate the molar mass.

When both components in a solution are volatile, the above formula applies to each component, provided the solution is ideal (see Fig. 8.2).

8.4.2. *Henry's law*

Henry's law relates the solubility of a gas in a liquid solvent to the gas pressure. An example is the solubility of oxygen in water. In Raoult's law, we start with a liquid and measure its vapor pressure as a function of the liquid mole fraction. In Henry's law, we start with a gas and measure its solubility in the liquid as the pressure is increased. In thermodynamic theory there is no difference between the two: they both are examples of equilibrium of a component between two phases.

There is, however, a difference in the type of data we get. In Henry's law studies, we can never reach a mole fraction of 1 for the

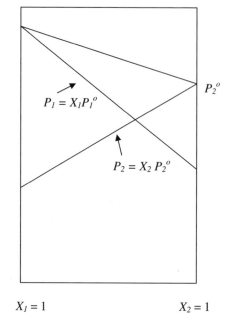

P_1^o

$P_1 = X_1 P_1^o$

P_2^o

$P_2 = X_2 P_2^o$

$X_1 = 1$ $X_2 = 1$

Fig. 8.2. Raoult's law for two volatile liquids. Vapor pressure of each liquid is proportional to its mole fraction.

soluble gas. In fact, we start with a mole fraction of 0 and integrating a log function from that initial value is not possible. However, this turns out not to be a problem. All our calculations involve differences between two states; in fact no one knows the absolute value of energy under any condition and it is only the energy differences that are manifested as various changes that we observe. Hence, for the integration of Eq. (8.20) we will set one of the limits as $X_2 = 1$ (the subscript indicates that the gas is the solute) and $P = k_H$ as the pressure at that hypothetical state of $X_2 = 1$.

$$\int_{k_H}^{P_2} d\ln P_2' = \int_1^{X_2} d\ln X_2' \qquad (8.23)$$

With this convention Henry's law takes the form

$$P_2 = k_H X_2 \qquad (8.24)$$

Figure 8.3 illustrates Henry's law.

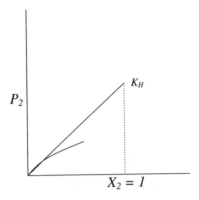

Fig. 8.3. Illustration of Henry's law. Clearly it is not possible to reach $X_2 = 1$ since that would mean no solvent. However, we can extrapolate the solubility to $X_2 = 1$ and use the hypothetical pressure of that state as the standard state. This is done for convenience only. As the lower curve shows, the solubility of a gas never reaches close to $X_2 = 1$.

8.4.3. *Solubility as a function of temperature*

In a typical chemistry experiment a solid compound (2) is dissolved in a liquid solvent (1) until some solid is left at the bottom. Let us find out how solubility varies with temperature. Temperature is a variable in both phases. Mole fraction varies in the liquid phase but not in the solid phase. The solution is thus saturated and the equilibrium is between the solid compound and its mole fraction in the solution ($X_2 = 1$ in the solid and $X_2 < 1$ in the solvent). With these conditions, Eqs. (8.8) and (8.11) combined give

$$-\frac{h_2(\text{dissolved})}{T^2}dT + Rd\ln X_2 = -\frac{h_2(\text{in solid form})}{T^2}dT \qquad (8.25)$$

From a thermodynamic point of view there is no difference between the solubility of a solid and its melting (fusion); in both cases a substance is undergoing phase transition. Therefore enthalpy difference in the above equation is taken to be the same as the enthalpy of fusion. With this understanding, integration of Eq. (8.25) gives

$$\ln X_2 = -\frac{\Delta h_2(\text{fusion})}{RT} + \text{Constant} \qquad (8.26)$$

where Δh_2 (fusion) $= h_2$ (in liquid phase) $- h_2$ (in the solid phase) is a positive quantity. There is an assumption here that enthalpies do not depend on temperature, which can be defended for small changes in temperature. Integration between limits gives

$$\ln \frac{X_1}{X_2} = \frac{\Delta h_A(\text{fusion})}{R} \left[\left(\frac{1}{T_2} \right) - \left(\frac{1}{T_1} \right) \right] \qquad (8.27)$$

Solubility increases with temperature since enthalpy change is positive. If the solution is truly ideal, we can determine the enthalpy of fusion by just measuring the solubility as a function of temperature. Since $\Delta S = \Delta H/T$, solubility increases if the substance has greater entropy in solution than in the solid form. Is this because of an increase in disorder or freedom?

Reality check

Equations (8.26) and (8.27) work only for a few systems, such as iodine and hexane. The ideal is a far cry from reality for most solid–liquid solubility equilibria. You are probably familiar with hot and cold packs whose function depends on solubility of salts in water or other liquids under non-ideal conditions. Nevertheless the above theory still enters into analysis of non-ideal systems, with modifications. Equation (8.26) is used like a rule for a small variation in temperature. Enthalpy is obtained by studying solubility at, say, two different temperatures, and then put to use for predicting solubility at a different temperature. Of course, enthalpy thus obtained does not have any relation to enthalpy of fusion, but is still meaningful as an empirical constant in a useful relation.

However, when there is a large amount of heat involved, as in the case of hot packs, it would be meaningless to use this theory. The heat generated is due to ionization and hydration, and solubility is a secondary phenomenon. The moral of this is that you have to know your chemistry before you can apply thermodynamic theory.

There is yet another reason that we are confining ourselves to ideal systems in this chapter. When you come across non-ideal systems, you will notice that the relevant equations look very much like the ones you see here. The difference is not in the theoretical expressions but in the variables that enter into the ersatz equation. The mole fraction gets replaced by activity and pressure by fugacity. Hence it is important that we first understand the ideal, however much it is removed from the real, before taking the next step.

Gases in liquids: Temperature dependence

The theory for gases in liquids is identical to the one above, except that entropy and enthalpy changes are between gas phase and liquid phase. If you replace Δh (fusion) in Eqs. (8.26) and (8.27) by $-\Delta h$ (vaporization) the equations apply to the solubility of gases. Since entropy of a dissolved gas is lower than its entropy in the gas phase, the solubility of gases decreases with temperature. If you take a glass of water from the refrigerator and leave it in a relatively warm room you will notice bubbles of air clinging to the sides after some time, indicating the lower solubility at higher temperature.

The theory sketched here should be modified in most real cases because entropy changes in the liquid solvent also affect the solubility.

8.5. Reaction Equilibrium

8.5.1. *Heterogeneous reactions*

Consider the reaction

$$CaCO_3(s) \leftrightarrow CaO(s) + CO_2(g)$$

The condition for equilibrium is the same as the one we used before: the chemical potential on the right-hand side must be equal to the chemical potential on the left. The mole fraction of CO_2 is unity irrespective of pressure and temperature, since it is the only component in the gas phase. This means that the only variable we have to consider is pressure at any given temperature. Therefore the condition for equilibrium (from Eqs. (8.10a) and (8.11)) is

$$\frac{v(CaCO_3)}{T}dP = \frac{v(CaO)}{T}dP + \frac{v(CO_2)}{T}dP \qquad (8.28)$$

To go beyond this point, we must integrate this expression. What should be the limits of integration? The accepted convention is to take the standard state as 1 bar at any temperature, meaning that we measure all changes relative to 1 bar pressure. Let the chemical potential in the standard state be designated by a superscript "o". Then we have from Eq. (8.28)

$$\mu^o(CaCO_3) + \int_1^P v(CaCO_3)dP' = \mu^o(CaO) + \int_1^P v(CaO)dP'$$

$$+ \mu^o(CO_2) + RT \int_1^P d\ln P' \qquad (8.29)$$

since T is a common factor. The ideal gas equation for carbon dioxide is used here. The change in volume of solids may be neglected, if pressure changes are not enormous. Hence the two integrals involving solids are very nearly zero and the above equation simplifies to

$$\mu^o(CO_2) - \mu^o(CaCO_3) - \mu^o(CaO) = -RT\ln\left(\frac{P}{1}\right) \qquad (8.30)$$

Usually "1" is omitted from the denominator. The left-hand side is the standard Gibbs energy change for the reaction and is designated by the symbol ΔG^o. Since chemical potential is molar Gibbs energy for each chemical, ΔG^o is the weighted sum of chemical potentials of the products minus the reactants.

$$\Delta G^o = \mu^o(CO_2) - \mu^o(CaCO_3) - \mu^o(CaO)$$

The expression in the brackets in Eq. (8.30) is a constant at given temperature. Hence it is called equilibrium constant and given the symbol K. With this convention Eq. (8.30) becomes

$$\Delta G^o = -RT\ln K$$

a form that has gained iconic status in chemical thermodynamics.

8.5.2. *Homogeneous reactions*

Let us consider ammonia synthesis, mentioned at the beginning of Part B chemical thermodynamics.

$$N_2(g) + 3H_2(g) \leftrightarrow 2NH_3(g)$$

We can use mole fractions or partial pressures as variables at constant temperature. The convention has been to use partial pressures for gas phase reactions since standard state is defined as the state at 1 bar pressure. Following the reasoning we used in the previous reaction, we find

$$2\mu^o(\text{NH}_3) - 3\mu^o(\text{H}_2) - \mu^o(\text{N}_2) = -RT\ln\frac{(P_{NH_3}/1)^2}{(P_{N_2}/1)(P_{H_2}/1)^3} \quad (8.31)$$

which is again written in the abbreviated form

$$\Delta G^o = -RT\ln K \quad (8.32)$$

8.5.3. *Calculation of equilibrium constants*

Thermodynamic theory is, however, useless if all we get from it is proof for the existence of the equilibrium constant. Most gas phase chemical reactions do not proceed without a catalyst; while collisions in the gas phase are frequent, they do not lead to product formation unless there is a third body to quench their energy. A catalytic surface on which molecules absorb provides a path for energy transfer and product formation. The catalysts, however, are not always selective and the products may further react to form unwanted side products. Hence a theory that can give even a ballpark estimate of the equilibrium constant is essential, and this is where thermodynamic theory provides valuable tools.

From the definition of Gibbs energy (Eq. (7.4)) we have that

$$\Delta G^o = \Delta H^o - T\Delta S^o \quad (8.33)$$

Entropy of each species in reaction is computed from heat capacity data as described earlier (Eq. (6.35)). Enthalpy is computed from heat capacity data and is also deduced by adding enthalpies of known reactions (Hess law). For molecules with spectroscopic data all three quantities in the above equation, as well as heat capacities, are calculated using the tools of statistical thermodynamics, while values of Gibbs energy, enthalpy and entropy culled from different sources and statistical thermodynamic calculations are listed in several places. Thus we can calculate ΔG^o for a reaction and estimate the equilibrium constant before designing a suitable apparatus.

8.5.4. Equilibrium constant as a function of temperature

It will simplify matters if we stick to a simple model reaction like

$$A(g) \leftrightarrow B(g)$$

We want to consider both temperature and pressure as variables. (Instead of pressure we could use mole fraction, but since standard states are defined as $P = 1$ bar, we will use pressure). From Eq. (8.8) and (8.11) we have

$$-\frac{h_A}{T^2}dT + \frac{v_A}{T}dP = -\frac{h_B}{T^2}dT + \frac{v_B}{T}dP \qquad (8.34)$$

With the help of the ideal gas law this equation is simplified to

$$d\ln\frac{P_B}{P_A} = -\frac{h_B - h_A}{RT^2}dT \qquad (8.35)$$

The ratio of pressures is the equilibrium constant. Indefinite integration gives

$$\ln K = -\frac{\Delta H}{RT} + \text{Constant}$$

Again we have an example of how enthalpy data allow us to predict the equilibrium constant at different temperatures.

    The main point of this chapter is to show the unity of the thermodynamic theory of equilibrium, but in order to do that I have glossed over several details. The overall conclusions are: pressure dependence of equilibrium is related to volumetric data or its simplifications based in an equation of state, and temperature dependence of equilibrium is related to enthalpies and heat capacities.

 Related material on the disc.

With CD-ROM

Chapter 9

Statistical Thermodynamics

In physical chemistry, statistical thermodynamics is a bridge between quantum theory and thermodynamics. The basic tenets of statistical thermodynamics are quite simple. Since macroscopic properties are properties of a large collection of molecules distributed among various quantum states, we should be able to compute the properties by averaging over quantum states of molecules. What we need then is a **distribution function** for the population of molecules in each quantum state.

9.1. The Boltzmann Distribution Function

The Boltzmann distribution function, in statistical thermodynamics, is the counterpart to the Helmholtz energy function in thermodynamics. When volume and temperature are held constant, Helmholtz energy, A, has a minimum value with respect to all non-equilibrium states at that volume and temperature. Under the same conditions the most probable distribution is the Boltzmann distribution.

The Helmholtz energy (Eq. (7.3)) and quantum energies are related by the equation

$$A/molecule = \sum_i p_i E_i - k_B T \sum_i p_i \ln p_i \qquad (9.1)$$

167

In this expression p_i is the probability that the molecule is in the ith state. The second term in the sum is entropy (Eq. (5.8)). Equation (9.1) is an expanded version of the equation:

$$A/molecule = u - Ts$$

We will use the symbol E for quantum energy and u for energy averaged over quantum energy levels, which corresponds to thermodynamic energy. Note that the theory introduced here applies *only* for ideal gases. It can be made more rigorous but we get the needed insight without going further. If intermolecular energy is not a factor, we can develop the theory for *a single molecule*, as Eq. (9.1) indicates, and scale it up for the number of molecules in the vessel. This is simpler than attempting to work with a larger system.

When volume and temperature are constant, the condition for A to be minimum (or maximum) is $dA = 0$. From Eq. (9.1) we have

$$dA = 0 = \sum dp_i(E_i - k_B T \ln p_i) \qquad (9.2)$$

Why do we need the volume to be constant? Otherwise we would have to consider E_i as a function of volume (see the "particle in a box" model in Section 1.5).

Since molecules leaving one energy state have to appear in some other state, dp_i are not independent. Hence we may be tempted to conclude from Eq. (9.2) that

$$p_i^U = \exp\left(-\frac{E_i}{k_B T}\right) \qquad (9.3)$$

This equation is correct up to a point. The probabilities are not normalized since the condition $\sum dp_i = 0$ does not guarantee that $\sum p_i = 1$. Hence the probability in Eq. (9.3) is "un-normalized" as indicated by the superscript. To normalize p_i^U, all we have to do is divide it by the total probability. While we are doing that, we can group all *states* with the same energy (degenerate states) into one *term*. Hence the final expression is

$$p_i = \frac{g_i \exp(-E_i/k_B T)}{\sum g_j \exp(-E_j/k_B T)} \qquad (9.4)$$

where g_j indicates the degeneracy of the jth *level*. The denominator in the above equation is called the **partition function** and is usually represented by q or z.

$$z = \sum g_j \exp(-E_j/k_B T) \qquad (9.5)$$

Note that the partition function is simply a sum of exponential terms. The German word for that, *zustandssumme* (sum over states), is very descriptive and hence we will use the symbol z for the partition function.

If we define

$$\beta \equiv \frac{1}{k_B T} \qquad (9.6)$$

Eq. (9.4) can be written in the simpler (and also more useful) form of

$$\boxed{p_j = \frac{g_j \exp(-\beta E_j)}{z}} \qquad (9.7)$$

This is the Boltzmann distribution function.

9.2. Thermodynamic Properties

Although the foundations of statistical thermodynamics are still being debated, its application to ideal gases is straightforward and the underlying concept is simple.

Step 1: Calculate the average energy from the equation

$$u = \sum p_j E_j \qquad (9.8)$$

and entropy from

$$s = k_B \sum p_j \ln p_j \qquad (9.9)$$

The probabilities, p_j, for the above equation are obtained from Eq. (9.7).

Step 2: Estimate thermodynamic properties from thus calculated u and s, and the partial derivatives in the previous chapters. For example, heat capacity at constant volume is a partial derivative of u with respect to T.

Now the details. From Eqs. (9.7) and (9.8) we get

$$u = \frac{\sum g_i E_i \exp(-\beta E_i)}{z} \tag{9.10}$$

Look at this expression carefully. The numerator is the differential of the term $\exp(-\beta E_i)$ with respect to β. Hence

$$u = -\frac{\dfrac{\partial}{\partial \beta} \sum g_i \exp(-\beta E_i)}{z} \tag{9.11}$$

Since the sum in the numerator is the partition function, another way of writing the above expression is

$$u = -\frac{1}{z}\frac{\partial z}{\partial \beta} = -\frac{\partial \ln z}{\partial \beta} \tag{9.12}$$

This is the starting point for the rest of the chapter. The following is the road map for our theory.

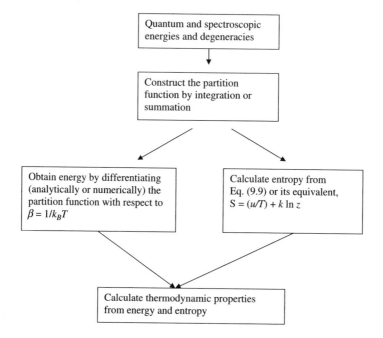

9.3. Partition Functions

As in Chapter 2 we will assume that the energy of a molecule is a sum of the energies of translations, rotations, vibrations and electronic states. We will then construct a partition function for each type of motion so that we can investigate its role in determining thermodynamic properties.

Electronic levels are so widely separated that only the lowest energy term ($E = 0$) is populated under normal conditions. Hence the exponential term becomes unity and the **electronic partition function**, z_e, is simply the degeneracy of the lowest term. For atomic hydrogen, $z_e = 2$ because of the two spin orientations. For the helium atom, $z_e = 1$ since the spins are paired. For most molecules, $z_e = 1$. The oxygen molecule, however, has a triplet ground term and hence $z_e = 3$.

9.3.1. *Translational partition function*

Translational energies are the energies of the particle in a box. They are so closely spaced as to form a continuum unless the box has nano-dimensions. Hence we can replace summation in the partition function by integration.

$$z_t = \int_o^\infty \exp(-\beta E_{n_x})dn_x \int_o^\infty \exp(-\beta E_{n_y})dn_y \int_o^\infty \exp(-\beta E_{n_z})dn_z \tag{9.13}$$

These are standard integrals and the final result is

$$z_t = C\beta^{-3/2} \tag{9.14}$$

where we have lumped all the distracting factors into a constant, C. From this equation and Eq. (9.12) we get

$$u_t = -\frac{1}{C\beta^{-3/2}}\frac{C\partial\beta^{-3/2}}{\partial\beta} = \frac{3}{2\beta} = \frac{3}{2}k_B T \tag{9.15}$$

As atoms only have translational degrees of freedom, this is all there is to their kinetic energy and temperature is a direct measure of it. From the above equation it is easy to see that constant volume heat capacity for atoms is

$$c_V \text{(translation)} = \frac{3}{2}k_B \text{ (per molecule)}$$

$$c_V \text{(translation)} = \frac{3}{2}R \text{ (per mole)} \tag{9.16}$$

Each of the three degrees of freedom contribute $(1/2)\ k_B T$ *for energy and* $(1/2)\ k_B$ *for heat capacity.* This result was first derived by Maxwell in 1860 but for nearly half a century, until Ramsey discovered the rare gases, there was no experimental verification. Before that all the known gases were either diatomic or polyatomic molecules.

The translational partition function without the abbreviation used before is

$$z_t = \frac{(2\pi m)^{3/2}V}{\beta^{3/2}h^3} = \frac{(2\pi m k_B T)^{3/2}V}{h^3} \tag{9.17}$$

Here m is the mass of the particle and V, the volume of the container. You may recall that $\frac{h}{\sqrt{2mE}}$ (example following Eq. (1.9)) is the de Broglie wavelength, λ. Since in this case $E = u = 3/2\ k_B T$, we have the following relation between the de Broglie wavelength and the partition function:

$$z_t \approx \frac{V}{\lambda^3} \tag{9.18}$$

This equation shows an interesting connection between the classical and quantum worlds. Since particles are really wave packets, λ^3 is the volume of the particle, approximately $1 \times 10^{-24}\ \text{cm}^3$. Hence in one liter ($1000\ \text{cm}^3$) there are about 1×10^{27} "cells" for the particle to sample. If you consider all the atoms in a liter at 1 atm pressure, you will find that the ratio of cells to molecules in a liter is approximately 4×10^5.

What happens if the number of molecules increases? We come to a point where the number of cells is too small for the Boltzmann statistics to work, even if intermolecular forces are negligible. If the temperature is very low, say near 0 K, according to quantum theory the wave packets (and hence what we call particles) become large and the ratio of cells to particles becomes small. Now we see the quantum effects of interference and superposition. Particles with half-integer spins (Fermions) repel each other while particles with integer spins (Bosons) clump together. This behavior is a consequence of the Pauli principle and manifests even when intermolecular forces vanish. We have to use Fermi–Dirac statistics for particles with half-integer spins and Bose–Einstein statistics for particles with integer spins.

9.3.2. *Rotational partition function: Diatomic molecules*

Rotational energies and their degeneracies for diatomic molecules are given in Chapter 2. Since rotational energies are discrete we should use a sum for the partition function. However, the calculations simplify if we assume continuity in rotational energies; an approximation justified *only* in the high temperature limit. The result of evaluating the partition function by integration is the following expression for the rotational partition function of diatomic molecules:

$$z_r = \frac{T}{\sigma \theta_r} \tag{9.19}$$

where

$$\theta_r = \frac{h^2}{8\pi^2 I k_B} \tag{9.20}$$

is called the rotational temperature. The symbol σ represents the symmetry number. It is 2 for homonuclear diatomic molecules and 1 for heteronuclear diatomic molecules. The simplified explanation for the symmetry number is that the two atoms in a homonuclear diatomic molecule are not distinguishable and hence we should count only half the number of states. Table 9.1 gives rotational and vibrational temperatures for a few molecules.

Approximating a sum by an integral, as we did to get Eq. (9.19), can be justified only if $T > \theta_r$. Thus for H_2, integration makes sense

Table 9.1. Rotational and vibrational temperatures for a few molecules.

Molecule	θ_r	θ_V
H_2	85.4 K	5755 K
O_2	2.06	2230
Cl_2	0.35	810
I_2	0.0054	310

only at high temperature. For I_2, on the other hand, it is a valid approximation at 25°C.

From Eqs. (9.11) and (9.19) we get the following energy for a molecule:

$$u_r = k_B T \tag{9.21}$$

and

$$c_V(\text{rotation}) = k_B \tag{9.22}$$

for heat capacity.

9.3.3. *Vibrational partition function*

Now we have to use summation to obtain the partition function. If we define vibrational temperature by the equation

$$\theta_V = \frac{h\nu}{k_B} \tag{9.23}$$

the partition function becomes

$$z_{v,o} = \sum_v \exp\frac{-\theta_V}{T} \tag{9.24}$$

The subscript "o" indicates that energies are measured from the zero point energy level. For now, let us set aside the detailed theory and proceed to the final result.

By following the procedure outlined in Section 9.2, we get the following expression for heat capacity at low temperature:

$$\frac{c_V(\text{vibrational})}{k_B} = \left(\frac{\theta_V}{T}\right)^2 \exp\left(-\frac{\theta_V}{T}\right) \qquad (9.25)$$

As T decreases, the exponential function falls off more rapidly and compensates for the rise in the quadratic term. Hence $c_V \rightarrow 0$ as $T \rightarrow 0$.

Even though we are focusing on diatomic molecules, we should note that the theory was first developed by Einstein in 1907 for atoms in monatomic crystals. Einstein's great insight was to realize that if energies of the atoms in the crystal are discrete, thermal energy of the surroundings cannot excite the atoms near 0 K and hence there will be no heat absorption. If the energy of atoms in a crystal is a continuous function, then they should be able to absorb heat from the surroundings even as we approach absolute zero. In that case $c_V > 0$, contrary to experiment. Figure 9.1 shows a plot of Einstein's function and some experimental results.

Vibrational heat capacity (dimensionless) for a diatomic molecule at high temperature is given by

$$\frac{c_V(\text{vib})}{k_B} = \frac{C_V(\text{vib})}{R} = 1 \qquad (9.26)$$

(The vibrational heat capacity is 3 for atoms in a crystal since they each have three vibrational degrees of freedom.)

9.4. Heat Capacities

Figure 9.2 shows heat capacities of a few gases. For monatomic gases, the experimental value for c_V is $3/2\,k_B$. This is in excellent agreement with our calculation for the translational partition function. The low temperature heat capacity of H_2 is barely above that of a monatomic gas. The reason for this becomes obvious when we look at Table 9.1. The rotational temperature for H_2 is too high for rotations to be excited at low temperature; only at 250 K and above can rotations contribute to heat capacity. At these temperatures $c_V = 3/2\,k_B$ (translations) $+ 1\,k_B$ (rotations), which is what has been

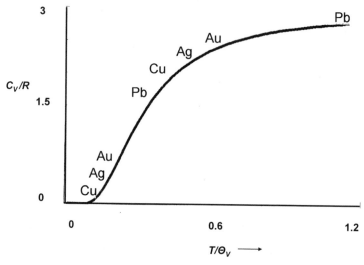

Fig. 9.1.   Heat capacity of monatomic crystals according to Einstein's theory (1907). The fit with experimental data (at 25 K and 100 K for four solids) is not particularly impressive. Nevertheless his work is a landmark in the development of quantum theory in that it shows only quantum theory can account for heat capacity vanishing near 0 K. By using a distribution for crystal frequencies, instead of one frequency as Einstein did, Debye developed an expression that fits the data much better.

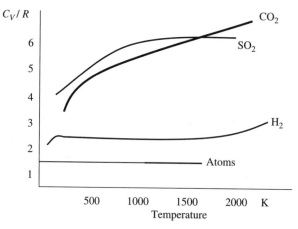

Fig. 9.2.   Heat capacity of atoms and the molecules H_2, SO_2 and CO_2. See discussion in Section 9.4.

observed. Due to the high vibrational temperature for H_2, vibrational excitation does not play any significant role below 1500 K.

Theory predicts $c_V = 6.5\,k_B$ and $6.0\,k_B$ for CO_2 and SO_2 respectively as shown below

	CO_2 (linear)	SO_2 (bent)
Translations	$3 \times 0.5\,k_B = 1.5\,k_B$	$3 \times 0.5\,k_B = 1.5\,k_B$
Rotations	$2 \times 0.5\,k_B = 1.0\,k_B$	$3 \times 0.5\,k_B = 1.5\,k_B$
Vibrations	$4 \times 1.0\,k_B = 4.0\,k_B$	$3 \times 1.0\,k_B = 3.0\,k_B$
Total	$6.5\,k_B$	$6.0\,k_B$

This agrees with what is observed at high temperature. However, in the intermediate temperature range, SO_2 has a higher heat capacity because its vibrations have lower frequencies and hence are more easily excited than CO_2 vibrations.

9.5. Chemical Equilibrium

The equilibrium constant is determined by the distribution of molecules in the energy states. Say the equilibrium constant for reaction $A \leftrightarrow B$ is 10; it simply means that there are ten times as many molecules in the quantum states of B than in A. Consider the energy level scheme in Fig. 9.3. After A and B are mixed there will be redistribution among the energy states (the real picture is a bit more complicated since intermolecular interaction changes the

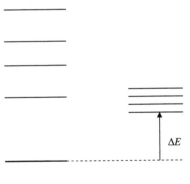

Energy states of A   Energy states of B

Fig. 9.3.   Energy states of A and B molecules. Both partition functions must be constructed from the same reference energy state (see Eqs. (9.29) and (9.30)).

energy levels but we need not consider that complication for gases at low pressure). The number of molecules in the ith energy state of A is given by

$$N_i^A = N \frac{\exp\left(-\beta E_i^A\right)}{Z} \tag{9.27}$$

where N is the total number of molecules and Z is the partition function for the *combined system*. The number of molecules in the A states is obtained by summation over all states of A. If you do this; the numerator in Eq. (9.27) becomes the partition function for A.

$$\frac{N^A}{N} = \frac{z^A}{Z} \tag{9.28}$$

Obviously we will have a similar expression for N^B. However, we need to take another factor into account. Energies must be measured from the same reference level. Hence the number of molecules in the B states is given by

$$\frac{N^B}{N} = \frac{z^B}{Z} e^{-\beta \Delta E} \tag{9.29}$$

(see Fig. 9.3).

From the above two equations it follows that the equilibrium constant, K, is given by

$$K = \frac{N^B}{N^A} = \frac{z^B}{z^A} \exp(-\beta \Delta E) \tag{9.30}$$

This argument can be readily extended to reactions with more than two chemicals.

9.5.1. *Example: Ionization equilibrium*

The simplest reaction that illustrates the ideas above is the ionization of atoms:

$$
\begin{array}{cccc}
\text{A} & \leftrightarrow & \text{A}^+ & + & \text{e}^- \\
1 - \alpha & & \alpha & & \alpha & \Leftarrow \text{degree of ionization} \\
(1 - \alpha)\text{P} & & \alpha P & & \alpha P & \Leftarrow \text{partial pressures} \quad (9.31)
\end{array}
$$

We need to consider only the translational partition functions, since that is the only degree of freedom available for these species. Further, translational partition functions are the same for A and A^+, since removing the electron does not change the mass of A appreciably. Hence

$$K_N = \frac{(2\pi m^e kT)^{3/2}}{h^2} e^{-\beta I} \tag{9.32}$$

Here I is the ionization energy; subscript N indicates we are expressing the equilibrium constant as a ratio of numbers. To see how this equation is used we have to do a bit of algebra. Since

$$P = \frac{N k_B T}{V} \tag{9.33}$$

we have

$$K_P = K_N \frac{k_B T}{V} \tag{9.34}$$

where K_P is the equilibrium constant with pressure as the variable. What is interesting here is the degree of ionization. It follows from Eq. (9.31) that the equilibrium constant is related to the degree of ionization by

$$K_P = \frac{\alpha^2 P}{1 - \alpha} \tag{9.35}$$

If the degree of ionization is small, the above equation simplifies to

$$\alpha = \sqrt{\frac{K_P}{P}} \tag{9.36}$$

Combining Eqs. (9.36), (9.34) and (9.32), we get

$$\alpha = \left(\frac{2\pi m^e}{h^2}\right)^{3/4} (kT)^{5/4} P^{-1/2} e^{-(\beta I/2)}$$

This equation was derived by Saha in 1920 and is considered a landmark publication in astrophysics. We need not go into the details of how the spectral lines of elements and ions will give information on degree of ionization. From such studies astronomers have been able to determine temperature and density (or pressure) of solar and stellar atmospheres.

This chapter has been restricted to ideal gases of atoms and diatomic molecules, in order to illustrate the principles of statistical thermodynamics with minimum effort. The main point is that once we have the partition function, calculation of thermodynamic properties is straightforward, even if it is laborious in some cases. In practice we do not need formulas for molecular energies. We can compute heat capacities, equilibrium constants and other thermodynamic properties just by manually entering temperature and spectroscopically determined energies into a computer program, even if they don't fit into a neat formula. This is of great value in gas phase equilibria, since lack of proper catalysts makes experiments difficult.

With CD-ROM

Related material on the disc.

Chapter 10

States of Matter

Solid, liquid and gaseous forms of materials are the three predominant states of matter. Since we observe a physical boundary between these forms, they are also called phases of matter. Let us see how molecular interactions determine the properties of gases and liquids. This brief chapter shows the connection between thermodynamics, equations of state and distribution functions — all the subjects we have explored before.

10.1. Non-Ideal Gases: Pair-Wise Interaction

In reality all gases are non-ideal, even though they obey the ideal gas law at low pressure and high temperature. There are two overlapping methods in the study of non-ideal gases. They depend on (1) equations of state and (2) potential energy functions for pair-wise interaction of molecules.

Equations of state are mathematical relations between P, V and T. As we have seen in Eqs. (6.20) and (6.22) with the van der Waals equation as an example, an equation of state greatly reduces the need for doing difficult experiments.

10.1.1. *London dispersion energy*

We have two frequently used mathematical expressions for pair-wise interaction of non-polar molecules. One was derived from quantum theory in 1930 by London, who called it the *dispersion effect*. Atoms and non-polar molecules have instantaneous dipole moments (freeze time and you will see that charge distribution is far from uniform). When averaged over time for measurement, the instantaneous dipole moments vanish. However, interaction between molecules propagates at the speed of light and they react with each other's instantaneous dipoles. The interaction between instantaneous dipoles also vanishes when averaged. London showed that an instantaneous dipole in one molecule induces a dipole in the neighboring molecules and that the interaction between **induced dipoles** leads to attractive potential between them. The following is the expression for the **induced dipole–induced dipole (dispersion)** energy between molecules A and B

$$E(\text{dispersion}) = -\frac{3}{2}\frac{\alpha_A\alpha_B}{(4\pi\varepsilon_0)^2 r^6}\frac{I_A I_B}{I_A + I_B} \tag{10.1}$$

In this equation α are the polarizabilities, I the ionization energies, r is the distance between the molecules and ε_0 is the permittivity of vacuum.

London potential falls off as the inverse sixth power of intermolecular distance. At large intermolecular distances *retardation effects* set in. In rough numbers electronic distribution changes with a frequency of $10^{18}\,\text{s}^{-1}$ and radiation travels with the speed of light, $3 \times 10^8\,\text{m\,s}^{-1}$. Hence beyond roughly 30 nm, molecules stop seeing each other and London forces become retarded.

10.1.2. *Lennard-Jones potential energy function*

We expect the attractive interaction between molecules to increase as the intermolecular distance decreases. At a certain minimum distance, however, repulsion should come into play; otherwise the molecules would coalesce into one single molecule. In 1924 Lennard-Jones suggested the following form for intermolecular potential energy, $\varphi(r)$:

$$\varphi(r) = 4\varepsilon\left[\left(\frac{\sigma}{r}\right)^{12} - \left(\frac{\sigma}{r}\right)^6\right] \tag{10.2}$$

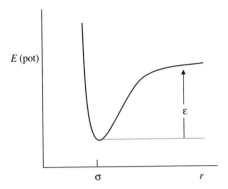

Fig. 10.1. The Lennard-Jones potential energy between two molecules. The contact distance is denoted by σ and the energy needed to separate the pair by ε. For helium, neon and argon pairs, σ is 258, 279 and 342 pm and (ε/k_B) is 10, 35.7 and 124 K, respectively.

The parameters in this equation are defined in Fig. 10.1. The attraction term has r^{-6} dependence, consistent with the dispersion energy. The repulsion term is simply the one that works; it is not derived from any fundamental theory. Lennard-Jones parameters are obtained by several methods. The equation finds wide use in statistical mechanics of non-ideal gases and liquids. A few Lennard-Jones constants for typical gases are listed in the caption to Fig. 10.1.

10.2. Continuity of the Fluid State

A phase boundary is commonly observed between a liquid and its vapor. As the pressure and temperature increase, the meniscus defining the boundary disappears at $T = T_c$ and $P = P_c$. T_c and P_c were called the critical temperature and pressure by Andrews (1869), who systematically studied the phenomenon. Figure 10.2 shows **isotherms** (constant temperature curves) from Andrews research.

When carbon dioxide gas at $21.5°C$ is compressed to 60 atm, a slight increase in pressure leads to an abrupt volume change and CO_2 becomes a liquid. If the same experiment is done at $31.4°C$, there will be no distinction between the gaseous and liquid states and therefore no abrupt change in the volume. At this critical state, a milky fluid with fluctuating globs of clouds (**critical opalescence**)

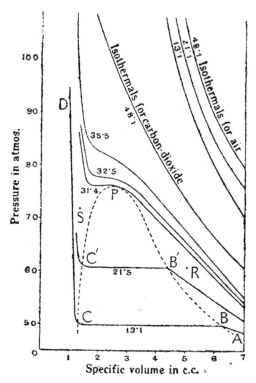

Fig. 10.2.   Isotherms for carbon dioxide. The figure is from Andrews' pioneering research (1869).

appears. Above this temperature there is no discontinuity in going from gas to liquid and we have to characterize the material as a fluid rather than a gas or a liquid. Research since Andrews' pioneering work showed that $T_c = 31.1°C$ and $P_c = 73.8$ bar.

At the critical point, $(\partial P/\partial V)_{T_c} = 0$. Hence isothermal compressibility,

$$\kappa = -\frac{1}{V}\left(\frac{\partial V}{\partial P}\right)_T$$

tends toward infinity as the fluid approaches critical state. This leads to a density gradient and fluctuations over a volume small enough for scattering of visible light. The result is the critical opalescence.

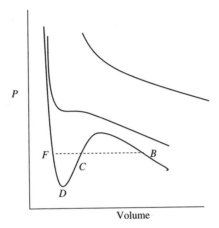

Fig. 10.3. The lower two curves show isotherm plots of the van der Waals
equation. The broken line in the lower curve represents the path for compression
near reversible conditions. The curve from B to F represents non-equilibrium
conditions. The middle curve is the critical isotherm. One of the successes of the
van der Waals equation is its prediction of the critical state. The top curve is
an isotherm above the critical point. There is no distinction between the gaseous
and liquid states above the critical point. The curves do not correspond to any
particular gas and the scale is distorted to bring out the features of the equation.

When he developed his celebrated equation, van der Waals was
interested in showing the continuity of gaseous and liquid states.
Figure 10.3 shows isotherms calculated from van der Waals' equation.
Since van der Waals' equation is cubic in volume, it has three roots.
This becomes clear if we expand his equation, Eq. (6.20), as shown
below.

$$V^3 - \left(b + \frac{RT}{P}\right) V^2 + a\frac{V}{P} - \frac{ba}{P} = 0 \qquad (10.3)$$

The three roots, B, C and F, are shown in Fig. 10.3. Point B
corresponds to the volume of the gas and F, to the volume of
the liquid. Parts of the BC segment correspond to supersaturated
vapor and the FD segment, to superheated liquid. These are meta-
stable states that suddenly, almost violently, change with small
perturbation.

The three roots of the van der Waals equation become identical at the critical point. Thus at critical point,

$$(V - V_c)^3 = 0 \qquad (10.4)$$

Expanding this equation gives

$$V^3 - 3V^2 V_c + 3V V_c^2 - V_c^3 = 0 \qquad (10.5)$$

By comparing the coefficients of V in Eqs. (10.5) and (10.3) (and doing a bit of algebra) we get

$$V_c = 3b; \qquad P_c = \frac{a}{27b^2}; \qquad T_c = \frac{8a}{27Rb} \qquad (10.6)$$

Thus the van der Waals equation can be used to predict the critical constants or, if critical constants are known, to deduce the van der Waals constants from them.

10.3. The Liquid State

10.3.1. *Pair correlation function*

Here we shall have to restrict our discussions to atoms in a liquid state. Suppose you can sit on an ideal gas atom and watch other atoms. You will find an atom in every position surrounding your atom, if you observe long enough. So the probability for a pair of atoms being at a certain distance r, called the **pair correlation function** and given the symbol $g_2(r)$, is constant beyond their distance of closest approach as shown in Fig. 10.4. The pair correlation function also goes by another name, **radial distribution function**.

The dotted line in Fig. 10.4 shows an estimate of how $g_2(r)$ for a non-ideal gas might look. Due to the dispersion forces between atoms there will be some clustering in a non-ideal gas, which increases $g_2(r)$ near the contact distance as shown in the figure. Atoms in a solid are fixed in space, so their pair correlation function will have finite value only at certain distances, but thermal agitation makes the peaks broaden. This is also shown in Fig. 10.4.

Solids have a long-range order and gases, no order. Liquids fall between the two; they have short-range order but the order

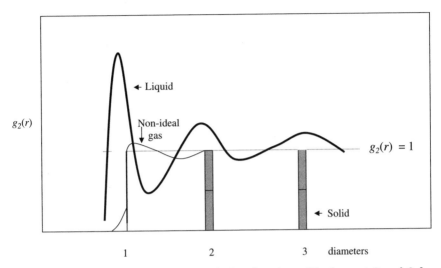

Fig. 10.4. A comparison of pair correlation functions. The bars at 2 and 3 for solid are just to compare its behavior with liquids and gases. Properties of a solid are directional and what is shown here does not correspond to any real solid.

trails off beyond the third neighbor as shown in Fig. 10.4. Bernal once aptly said that a solid is a pile of "bricks" and the liquid, a heap. The pair correlation function of liquids is measured by x-ray scattering techniques. Measurements at different temperatures show pair correlation function for liquids becoming diffuse and approaching the shape for non-ideal gases.

From pair correlation function we get the radial distribution function, $\rho(r)$:

$$\rho(r) = 4\pi r^2 \frac{N}{V} g_2(r) \tag{10.7}$$

Pair correlation function gives probability along a particular axis. Radial distribution function, on the other hand, counts all pairs at a distance r in all directions. The number of atoms surrounding the chosen atom, called the coordination number (Z), is given by

$$Z = \int_0^{r(\min)} \rho(r) dr \tag{10.8}$$

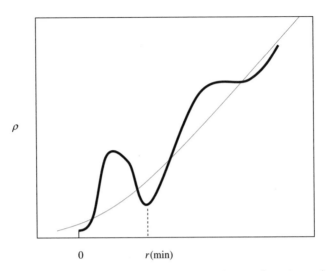

Fig. 10.5. The broken line shows the radial distribution function, $\rho(r)$, for an ideal gas. The solid line shows the same function for a liquid. The area between 0 and $r(\min)$ gives the number of atoms in the first coordination sphere.

The volume of a spherical shell between r and $r + dr$ is $4\pi r^2 dr$. The integral is simply the volume of the first coordination sphere, the volume containing the immediate neighbors, multiplied by the probability of finding atoms in that volume (see Fig. 10.5). The coordination number is usually between six and ten.

If the total inter-atomic energy of a liquid is a sum of pair-wise interactions, potential energy of a mole of atoms is given by

$$E(\text{pot})/\text{mol} = \frac{2\pi N_A^2}{V} \int_0^\infty g_2(r)\phi(r)r^2 dr \qquad (10.9)$$

This equation says: multiply probability at distance r with potential energy at r and integrate over the volume element $4\pi r^2 dr$ to include all pairs in volume V. Use 2 instead of 4 to avoid counting the A–B and B–A pairs as different. N_A is the Avogadro number.

The heat capacity of liquids is higher than that of gases. Heat capacity of O_2 shown in Fig. 6.4 is a good example. Turning to atoms, we have noted before that $C_V = (3/2)R$ in the gas phase (3 translational degrees of freedom) and $3R$ in the solid phase (3 vibrational degrees of freedom). For aluminum in the liquid phase, heat capacity is $3.53\,R$. A qualitative explanation for this is that the pair correlation function, and hence energy, increases with temperature. This adds to the heat capacity of individual atoms and molecules.

My aim in this chapter has been to give a glimpse into the nature of non-specific intermolecular forces. From an understanding of such forces we developed a scenario for continuity of gaseous and liquid states and the nature of the liquid state. In particular we were able to explain short-range order in liquids and why the liquid of a substance has higher heat capacity than its gaseous or solid forms. A comprehensive study must include specific interactions between molecules as well as the dependence of energy on intermolecular distance.

 Related material on the disc.

With CD-ROM

Part C

Kinetics

When Haber discovered a method of making ammonia from nitrogen and hydrogen, his findings had the potential to change world history. Since both explosives and agrichemicals are derived from ammonia, the First World War was prolonged with its concommitant misery, and agricultural productivity went up to feed the hungry humanity and fuel population growth. When Nobel developed dynamite and gave us a technique for handling explosives safely, the course of world history again changed. Development of railways through mountainous areas, safe mining and building of dams became practical.

Haber had to find a way to coax the reaction to go at an appreciable rate and Nobel had to find a way to slow down the reaction. We see from this that understanding and controlling the rates of reactions is an important part of chemistry. Thermodynamics is a reliable guide for predicting the direction of change and the yield of products, but it cannot help us infer the speed at which reactions happen. In the next two chapters we will explore the core ideas in kinetics.

Chapter 11

The Kinetic Molecular Theory of Gases

11.1. Distribution of Speeds   11.3. Transport

11.2. Collisions

The kinetic theory of translational motion in the gas phase has charming simplicity and elegant imagery. The theory relates pressure to molecular collisions with the surface of the container, and temperature to molecular speeds. From the **distribution of speeds** we find the frequency of collisions between molecules, rates for transport of heat, momentum and mass. Kinetic theory also provides the background for understanding rates of chemical change.

11.1. Distribution of Speeds

The Maxwell–Boltzmann distribution function

$$f(c)dc = 4\pi \left(\frac{m}{2\pi k_B T} \right)^{3/2} c^2 \exp \left(\frac{-mc^2}{2k_B T} \right) dc \qquad (11.1)$$

is the starting point for the rest of the material in this chapter. You are now familiar with probability density (Chapter 1) and distribution functions (Chapter 9). So instead of deriving Eq. (11.1), We can examine its special features. The exponent on the right-hand side of the equation is the ratio of kinetic energy to thermal energy, $k_B T$; this is identical to the Boltzmann exponential term. Instead of the regular space, think of a "space" with three perpendicular axes representing the three components of velocity. The term $4\pi c^2 dc$ is the volume in a spherical shell of thickness dc at a distance c

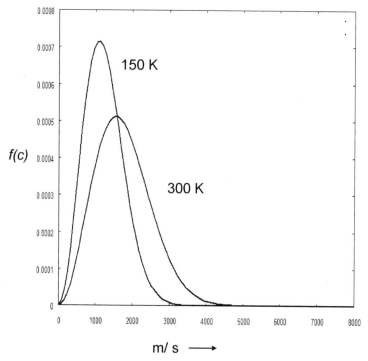

Fig. 11.1. The Maxwell distribution function for hydrogen molecule at two different temperatures.

in the velocity space. This is very much like the degeneracy term we have seen in discrete distribution. $f(c)$ is the probability density $[m^{-1}s]$ function for speeds. The equation has two opposing terms: as c increases the exponential term decreases with energy, but the number of available states increases. Hence the function rises initially, but decreases as c becomes large. Figure 11.1 shows the distribution function at two temperatures.

11.1.1. *Variety of speeds*

There are several measures for speed. Kinetic energy and temperature are related by (Eq. (9.15))

$$\frac{1}{2}mc^2 = \frac{3}{2}k_BT$$

Hence **root-mean-square speed (RMS)** is given by

$$c_{RMS} = \sqrt{\frac{3k_BT}{m}} \tag{11.2}$$

The word "root-mean-square" should be read backward: square first, take the average next and finally take the square root of the average.

Most probable speed corresponds to the maximum in the distribution function and is obtained from the equation

$$\left(\frac{df(c)}{dc}\right) = 0 \tag{11.3}$$

The result of differentiation is

$$c_{mp} = \sqrt{\frac{2k_BT}{m}} \tag{11.4}$$

We will find average speed useful in estimating the frequency of collisions.

$$\langle c \rangle = \sqrt{\frac{8k_BT}{\pi m}} \tag{11.5}$$

The average speed is obtained from the following integral:

$$\int_0^\infty f(c)c\,dc \tag{11.6}$$

From the above formulas we get the following values for the different measures of speed of N_2 at 21°C:

$$c_{mp} = 418\,\mathrm{m\,s}^{-1} \qquad \langle c \rangle = 471\,\mathrm{m\,s}^{-1} \qquad c_{RMS} = 512\,\mathrm{m\,s}^{-1}$$

It is worth noting that these speeds are in the 1500 to 1800 km hour^{-1} range.

Figure 11.2 shows an experimental arrangement for determining the distribution of speeds.

11.2. Collisions

Molecules "communicate" with each other through collisions. Diffusion and rates of heat and momentum transfer depend on frequency

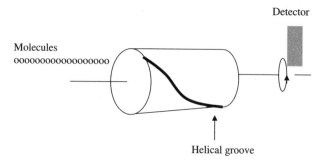

Detector

Molecules
ooooooooooooooooooo

Helical groove

Fig. 11.2. An apparatus for measuring molecular speeds. Only molecules whose linear velocity matches the angular velocity of the cylinder pass through and are detected.

of collisions. In the simple theory presented here, molecules are considered hard spheres. This would appear to be ignoring the fact that molecules are not spheres and they change in size while vibrating. Nevertheless, the hard sphere approximation works at ordinary pressure and temperature because molecules take all possible orientations, when averaged over a collection and over time, and act therefore like hard spheres.

11.2.1. *Frequency of collisions*

How often does a molecule experience collisions? We will use dimensional analysis to get the formula since it gives insight into the process. Let $Z_1 [\mathrm{s}^{-1}]$ stand for the number of collisions a molecule experiences during one second. We expect it to depend on the cross-section of the molecule $[\mathrm{m}^2]$, its average speed $[\mathrm{m\,s}^{-1}]$ and the number of other molecules, N, in the volume, $V [\mathrm{m}^3]$, it sweeps out. It is reasonable to assume that a collision happens when two molecules touch. That means when they collide their centers of mass are one diameter, d, apart. We will therefore use πd^2 as the collision cross-section. Thus we obtain

$$Z_1 = \sqrt{2}(\pi d^2)\langle c \rangle \left(\frac{N}{V}\right) \qquad [s^{-1}]$$
$$[\mathrm{m}^2][\mathrm{m\,s}^{-1}][\mathrm{m}^{-3}]$$

$$(11.7)$$

Since all molecules are in motion, we have to consider their relative speeds. For like molecules, the reduced mass is one half the mass of one molecule. This leads to the factor $\sqrt{2}$ if mass of the molecule, instead of reduced mass, is used to calculate $\langle c \rangle$.

Taking a reasonable value of $370 \times 10^{-12} \, \text{m}^2$ for the cross-section in the above formula, we get $Z_1 = 7.2 \times 10^9 \, \text{s}^{-1}$ for N_2 at 1 atm. Under these conditions a N_2 molecule experiences, on the average, a collision every 1.4 nanosecond, which alters its energy and momentum.

11.2.2. *Collision number*

Let Z_{12} be the number of collisions among all like molecules in a volume, V. This is obtained by multiplying Z_1 by the number of molecules and dividing it by 2. The division ensures that collision between a pair of molecules is not counted twice (A colliding with B is the same as B colliding with A). The formula then is

$$Z_{12} = \frac{1}{\sqrt{2}} \left(\frac{N}{V} \right)^2 \pi d^2 \langle c \rangle \quad [\text{m}^3 \, \text{s}^{-1}] \tag{11.8}$$

For N_2 at 21°C and 1 atm pressure, $Z_{12} = 8.9 \times 10^{34} \, [\text{m}^{-3} \, \text{s}^{-1}]$.

11.2.3. *Mean free path*

Mean free path, $\lambda[\text{m}]$, is the average distance covered by a molecule between collisions. It is equal to the distance traveled in a second divided by the number of collisions during a second.

$$\lambda = \frac{\langle c \rangle}{Z_1} \frac{[\text{m s}^{-1}]}{[\text{s}^{-1}]}$$

$$= \frac{1}{\sqrt{2}(N/V)\pi d^2} \quad [\text{m}] \tag{11.9a}$$

Since $PV = Nk_BT$, an alternative formula for mean free path is

$$\boxed{\lambda = \frac{1}{\sqrt{2}} \frac{k_BT}{P} \frac{1}{\pi d^2}} \tag{11.9b}$$

For N_2 at 21°C and 1 atm, $\lambda = 66 \, \text{nm}$. This is about 200 molecular diameters.

11.3. Transport

Molecules diffuse from a high to a low-concentration region. Heat flows from a hotter to a colder region. Momentum is transferred between fast and slow-moving layers, making the flow viscous. These are three examples of transport processes. In each case there is a gradient — concentration, temperature or momentum — and a rate at which transport takes place. The mathematical equations that relate the gradient to rate are called **phenomenological equations**.

11.3.1. *Diffusion*

The phenomenological equation for diffusion in one dimension is

$$J_N = -D\frac{d(N/V)}{dx} \tag{11.10}$$

In this equation, (N/V) is the number of molecules in a unit volume $[\mathrm{m}^{-3}]$, J_N is the number of molecules crossing a unit area in unit time $[\mathrm{m}^{-2}\mathrm{s}^{-1}]$ and D is called the diffusion coefficient. It can be readily verified that D must have the units $[\mathrm{m}^2\mathrm{s}^{-1}]$. The negative sign in Eq. (11.10) indicates that net diffusion takes place from a high to a low-concentration region.

The word "phenomenological" implies that we are just describing a phenomenon through an equation with no prior commitment to any molecular model. However, kinetic theory and thermodynamics between them give a mechanism for diffusion.

According to thermodynamics, the change is always toward equilibrium, even if the rate is minimal. Equilibrium is attained only when chemical potential is uniform throughout the system. Hence molecules drift until concentration and chemical potential are the same in every part of the vessel. According to kinetic theory, there is a flow in both directions which is retarded by collisions. However, the number of molecules crossing a plane from a higher to a lower concentration region is greater than the number drifting in the opposite direction, leading to net flow from high to low concentration.

From dimensional analysis we see immediately that the diffusion coefficient is proportional to the product of mean free path and average speed. When intermolecular forces are vanishingly small, the diffusion coefficient for a gas is given by

$$D = \frac{1}{3}\langle c \rangle \lambda \tag{11.11}$$

Dimensional analysis gives the above formula except for the factor $\frac{1}{3}$. If you want a justification for the numerical factor, keep in mind that only a component of the average speed is relevant to one-dimensional diffusion.

Everything said so far about diffusion only applies to one-dimensional self-diffusion. Nevertheless it gives a glimpse into the general characteristics of diffusion, which are listed below.

1. Since average speed depends on $T^{1/2}$ and mean free path on T, diffusion coefficients vary as a function of $T^{3/2}$.
2. "Diffusion speed" and "diffusion velocity" are meaningless phrases. Suppose you have a layer of high concentration in the middle of a tube (Fig. 11.3). Molecules from this layer diffuse to each side with equal probability. Thus the average position of all the diffusing molecules remains $x = 0$. The mean squared

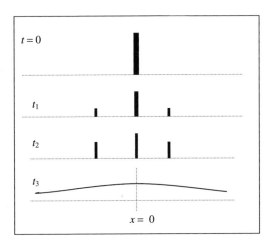

Fig. 11.3. Illustration of one-dimensional diffusion. From the center layer at $x = 0$, molecules diffuse to each side with equal probability. Average position of all molecules remains at $x = 0$. The mean square displacement, however, is non-zero.

displacement, on the other hand, depends on time. From just checking the dimensions we see that

$$\langle x^2(t) \rangle = \text{constant } Dt \qquad (11.12)$$

Detailed theory shows that the constant is 2 for one-dimensional diffusion, and 4 and 6 for two- and three-dimensional diffusion respectively.

3. Self-diffusion coefficients for H_2 and O_2 are 1.28 and 0.187 cm^2s^{-1}. The larger value for H_2 is expected since it has greater average speed and smaller diameter than O_2. Self-diffusion coefficients are measured by monitoring the movements of isotopes.

11.3.2. *Viscosity*

We will only consider gases here, but this can also be helpful in understanding the behavior of liquids and solutions. Consider the flow of gas between two parallel plates. The molecules next to the surface of the plates hardly move (Fig. 11.4) and they retard the flow of the layers next to them. In a **laminar flow** each layer flows at a different speed, the middle layer being the fastest. Molecules jump from layer to layer "carrying" their momentum with them. This leads to retardation of fast layers and acceleration of the slow layers. The phenomenological equation for flow in the x-direction is

$$\frac{F}{A} = -\eta \frac{dc_x}{dy} \qquad (11.13)$$

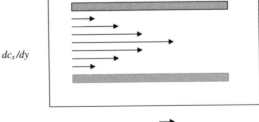

dc_x/dy

$x \longrightarrow$

Fig. 11.4. Laminar flow. The layers close to the plates retard the flow of the layers next to them.

In this equation $F[\text{kg m s}^{-2}]$ is the shearing force between adjacent layers and $A[\text{m}^2]$ is the area between them. The velocity gradient is given by (dc_x/dy); η, the proportionality constant, is called the coefficient of viscosity or simply viscosity. Its dimensions are kg m^{-1}s^{-1} and the SI unit is Pascal second (Pa s). The older unit $(1\,\text{poise} = 0.1\,\text{Pa s})$ is still used in the literature.

Since viscosity is a result of molecules jumping between the layers, η depends on mass of the molecules, their average speed, the average length of the jump (mean free path) and the number of molecules per unit volume. Dimensional analysis gives the formula

$$\eta = \frac{1}{3}m[\text{kg}]\langle c\rangle[\text{m s}^{-1}]\lambda[\text{m}]\frac{N}{V}[\text{m}^{-3}] \qquad (11.14)$$

The factor of $\frac{1}{3}$ comes from a more detailed theory. Typical values for viscosity are $18.5[\mu\text{Pa s}]$ for helium and $2.97[\mu\text{Pa s}]$ for neon.

The main ideas in this chapter concern the speed of molecules, the collisions they experience and the mean distance they travel before a collision changes their momentum. This is the first chapter that brings dynamics into focus and thus forms the basis for understanding chemical kinetics described in the next chapter.

 Related material on the disc.

With CD-ROM

Chapter 12

Chemical Kinetics

Chemical kinetics is the study of rates of chemical processes and the molecular mechanisms that control these rates. At the risk of sounding trite, one could even say that rate determines fate. This is particularly obvious in life processes, where the rates of various biochemical processes have to follow the Goldilocks principle and be just right. Either too fast or too slow leads to disease or death.

The essential datum in kinetic studies is the time dependence of the concentration of one of the species in the reaction vessel. **Rate** is defined by the equation

$$\text{Rate} = \frac{d[\text{A}]}{dt}$$

where [A] is the concentration of the chemical monitored, and may be positive or negative, depending on the choice of chemical. Generally speaking, rate depends on concentrations of all species present in the vessel, including the one used to define the rate. The relation between rate and the concentration of the chemicals is the **rate law**. Rate laws have the form

$$\text{Rate} = k[\text{A}]^a[\text{B}]^b \cdots$$

202

where the superscripts are the exponents and the bracketed quantities are the concentrations of the chemicals. The **rate constant**, k, is independent of the concentrations but may depend on pressure and temperature. The rate laws and rate constants must be determined experimentally, since the exponents in the above equation do not usually correspond to the mole numbers in the balanced reaction.

Rate laws are phenomenological equations in the sense that they only represent the data collected and do not indicate the **mechanism**; that is the intermediate reactions that lead to the final product. Let us consider a few specific examples.

12.1. Unimolecular Reactions in the Gas Phase

In these reactions, the reactant is a single chemical species. It seems to undergo *slow* but spontaneous change. Examples are dissociation of ethyl iodide,

$$C_2H_5I \rightarrow C_2H_4 + HI$$

and isomerization of methylisocyanide.

$$CH_3NC \rightarrow CH_3CN$$

The mechanism of these reactions puzzled chemists for decades. If the reactant were stable, it should not dissociate or isomerize. If it were not stable, why did it linger for a long time? The puzzle was finally solved by Lindemann in 1922. In the mechanism he suggested, the reactant is activated by collisions with other molecules in the container. The activated molecule either dissociates or becomes deactivated by another collision. If A is the reactant, the suggested mechanism is represented by

$A + M \rightarrow A^*$	rate constant $= k_1$	Activation
$A^* + M \rightarrow A$	rate constant $= k_2$	Deactivation
$A^* \rightarrow P(\text{product})$	rate constant $= k_3$	Product formation

M is any molecule that activates and deactivates molecule A through a collision. It could be another A molecule or an inert gas introduced into the reaction vessel. Since P is produced only in the last step, the rate for its production is

$$\frac{d[\mathrm{P}]}{dt} = k_3[\mathrm{A}^*] \qquad (12.1)$$

This equation is not of much use since there is no way of finding the concentration of activated molecules. However, there is a way to proceed further.

Low Pressure Limit

Consider what happens at low pressure. Since collisions are infrequent, there is a low probability of activation happening. Of course, that means that deactivation has an even lower probability; the activated molecule will stay long enough for the third step. Hence at *low pressure* we have

$$\mathrm{A} + \mathrm{M} \rightarrow \mathrm{A}^* \rightarrow \mathrm{P}$$

Once A^* is formed, it stays until further reaction takes place. This means that the whole process is controlled by the first step, giving the following rate law:

$$\frac{d[\mathrm{P}]}{dt} = k_1[\mathrm{A}][\mathrm{M}] \qquad (12.2)$$

This is a **second-order rate law** because rate is proportional to the product of two concentrations.

In this reaction A may be activated by another A molecule. So what happens if we only have A in the vessel and we monitor decay of A instead of growth of P? Equation (12.2) then becomes

$$-\frac{d[\mathrm{A}]}{dt} = k_1[\mathrm{A}]^2 \qquad (12.3)$$

High Pressure Limit

Under high pressure, collisions are frequent and both activation and deactivation steps are equally probable. The first two steps add to give an equilibrium concentration of A^*.

$$A \leftrightarrow A^* \quad \text{and} \quad K = \frac{k_1}{k_2} \tag{12.4}$$

Hence the rate law becomes

$$-\frac{d[A]}{dt} = \frac{k_1}{k_2} k_3 [A] \tag{12.5}$$

This is a **first-order rate law** since the rate is dependent on the concentration of one species. The rate law gives the false impression that molecules are undergoing change spontaneously.

Intermediate Pressure

To develop a more general rate law we use **steady state concentrations** for **intermediate species**. If the rate of production equals rate of reaction for a particular intermediate species, then its concentration does not change with time until the overall reaction is completed. We describe the intermediate species as being in a steady state.

Examine the three steps in the reaction; you will see that A^* is the only candidate for the steady state hypothesis, since it is produced in step 1 and destroyed in steps 2 and 3. Hence from step 1

$$\frac{d[A^*]}{dt} = k_1 [A][M] \tag{12.6}$$

and from steps 2 and 3

$$\frac{d[A^*]}{dt} = -k_2 [A^*][M] - k_3 [A^*] \tag{12.7}$$

The sum of these two rates must be zero at steady state

$$k_1 [A][M] - k_2 [A^*][M] - k_3 [A^*] = 0 \tag{12.8}$$

From this we have

$$[A^*] = \frac{k_1 [A][M]}{k_2 [M] + k_3} \tag{12.9}$$

Combining Eqs. (12.1) and (12.9) we get

$$-\frac{d[A]}{dt} = \frac{k_3 k_1 [A][M]}{k_2[M] + k_3} \tag{12.10}$$

This is the general rate law, valid over a wide pressure range, if the steady state approximation holds. Let us see how it compares with the previous results.

At high pressure $k_2[M] > k_3$ in the denominator and Eq. (12.10) reduces to

$$-\frac{d[A]}{dt} = \frac{k_3 k_1 [A][M]}{k_2[M]} = \frac{k_3 k_1}{k_2}[A]$$

which is identical to Eq. (12.5).

At low pressure $k_3 > k_2[M]$ and Eq. (12.10) becomes

$$\text{Rate} = k_1[A][M]$$

which is the same as Eq. (12.2). Figure 12.1 shows how the rate changes with [M].

There is no obvious way to predict the mechanism for a reaction; we simply have to use our knowledge of chemistry and work within the confines of the kinetic theory. The three basic ideas of kinetic theory are that (1) collisions involving three or four molecules are improbable (2) frequency of two body collisions depends on

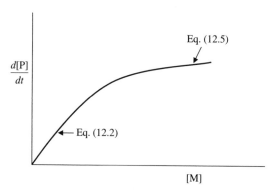

Fig. 12.1. Kinetics of a unimolecular reaction. At low concentrations of the activating species, M, rate is linear with [M]. At high concentrations of [M] the rate is independent of M.

pressure and (3) the concentrations of intermediate species have to be surmised based on the steady state or equilibrium hypothesis.

12.2. Testing Rate Laws

12.2.1. *First-order rate law*

A common method for verifying the rate law is to test it in its integral form. A first-order rate law for decay of a species can be brought to the form

$$-\frac{d[A]}{[A]} = kdt \tag{12.11}$$

(see Eq. (12.5) for comparison). The dimension for the first-order rate constant is $[s^{-1}]$. Integration of the above expression gives

$$[A] = [A_0]e^{-kt} \tag{12.12}$$

which can also be written as

$$\ln\frac{[A]}{[A_0]} = -kt \tag{12.13}$$

In these equations $[A_0]$ is the concentration at $t = 0$. It is clear from the above equation that if the logarithm of concentration is plotted against time, we should get a straight line with the slope $-k$. Figure 12.2 shows characteristic exponential decay of first-order processes.

Half-Life

Half-life, $t_{1/2}$, is the time it takes for the reactant concentration to reach half of the starting value. Note that in Eq. (12.13) when $t = t_{1/2}$, $[A] = 1/2[A_0]$, regardless of when we start the clock. Hence for a first order reaction

$$t_{1/2} = -\frac{1}{k}\ln\left(\frac{1}{2}\right) = \frac{\ln 2}{k} = \frac{0.693}{k} \tag{12.14}$$

It is a characteristic of first-order reactions that whatever the concentration is now, half of it will remain after one half-life. Radioactive

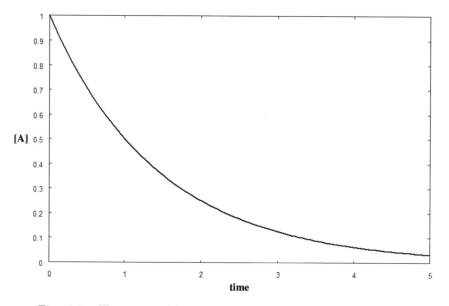

Fig. 12.2.   Illustration of first-order decay. Time is in units of half-life.

decay is probably the most dramatic example of first-order decay. One-sixteenth of an undesirable isotope that decays by half in 100 years will still be present 400 years later.

The first-order rate law also appears in growth processes. In growth kinetics, the sign of the exponent in Eq. (12.12) becomes positive. The equivalent of half-life now is doubling time. You may have heard of "the rule of 70," used in popular parlance. It is an adaptation of Eq. (12.14) for doubling time by expressing the rate constant for growth as a percentage and rounding off the numerator to 70.

12.2.2. *Second-order rate law*

If the second-order rate law is in the form

$$\frac{-d[A]}{dt} = k[A]^2 \qquad (12.15)$$

integration between the limits $t' = 0$ and $t' = t$ gives

$$\frac{1}{[A]} - \frac{1}{[A_0]} = kt \qquad (12.16)$$

Plotting $1/[A]$ as a function of time gives the rate constant. The dimensions for a second order rate constant are $[m^3\,mol^{-1}\,s^{-1}]$ but the units $[L\,mol^{-1}s^{-1} = M^{-1}s^{-1}]$ are used in the literature.

From Eq. (12.16) we get the half-life of a second-order reaction

$$t_{1/2} = \frac{1}{k[A_0]} \tag{12.17}$$

12.3. Bimolecular Reactions in Solution

The following represents the mechanism for a bimolecular reaction in solution:

$$A + B \rightleftharpoons (AB) \rightarrow \text{products} \tag{12.18}$$

The first step is diffusion of the two species, A and B, into a **solvent cage** to form a loose complex. That is the only step in a **diffusion-controlled reaction.** The reactants transform to products the moment they find themselves in proximity. In **activation-controlled reactions**, the AB complex forms after multiple collisions between A and B and remains in equilibrium with A and B. The third step is slow because molecular rearrangement is needed for the reaction to proceed. This section will be limited to diffusion-controlled reactions.

The rate law for diffusion-controlled reactions is

$$\text{Rate} = k_D[A][B]$$

Smoluchowski derived the following rate constant expression for the diffusion-controlled coagulation of colloidal gold in 1917:

$$k_D = 4\pi(D_A + D_B)(r_A + r_B)N_A \tag{12.19}$$

where the Avogadro number, sum of the two molecular radii and the sum of their diffusion coefficients appear on the right-hand side of the equation. This equation also applies to bimolecular reactions in solution if their rate is controlled by diffusion.

Diffusion-controlled reactions are some of the fastest reactions investigated by chemists. The recombination of H^+ and OH^- in aqueous medium has a rate constant of $1.3 \times 10^{11}[M^{-1}s^{-1}]$.

The technique used to measure this rate constant is described in Section 12.4.

If we assume that solute molecules are spherical with a radius r, an assumption justified when averaged over the time and the number of molecules, the diffusion coefficient of the *solute* (D) is related to the viscosity of *solvent* $[\eta]$.

$$D = \frac{k_B T}{6\pi\eta r} \tag{12.20}$$

This equation is known as the Stokes–Einstein equation.

By substituting Eq. (12.20) into Eq. (12.19) we get the following expression for a bimolecular reaction:

$$k = \frac{2k_B T}{3\eta}(r_A + r_B)\left(\frac{1}{r_A} + \frac{1}{r_B}\right) \tag{12.21}$$

It is easy to check that for molecules with comparable radii $(r_A \approx r_B)$ this equation reduces to

$$k = \frac{8k_B T}{3\eta} \tag{12.22}$$

We see from this equation that the rate constant for diffusion controlled reactions depends only on the solvent.

12.4. Relaxation Method for Determining Rate Constants

In this technique, we allow the reaction to come to equilibrium and then perturb it by sudden change in temperature or hydrostatic pressure. The system cannot reach equilibrium immediately under the new conditions. The approach to equilibrium (relaxation, as it is called) is a first-order process, if the perturbation is small, and from the measurement of the relaxation rate constant we can deduce the forward and reverse kinetic rate constants. Let us consider the ionization of water as an example.

$$H_2O \underset{k_r}{\overset{k_f}{\rightleftarrows}} H^+ + OH^-$$

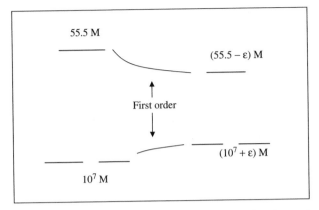

55.5 M

$(55.5 - \varepsilon)$ M

First order

$(10^7 + \varepsilon)$ M

10^7 M

Fig. 12.3. When the temperature of water is suddenly changed, ionization equilibrium is perturbed. Equilibrium at the final temperature is reached through a first-order process. Measuring relaxation time for this process allows us to determine very fast rates for ionization and association.

Figure 12.3 gives the concentrations of the species in equilibrium at two different temperatures. The subscript "0" indicates the initial concentrations and ε, the change in the concentrations. We measure ε as a function of time in relaxation techniques. The rate equation for ε is

$$\frac{d\varepsilon}{dt} = k_f \left([H_2O]_0 - \varepsilon\right) - k_r \left([H^+]_0 + \varepsilon\right) \left([OH^-]_0 + \varepsilon\right) \qquad (12.23)$$

There are two legitimate simplifications for this equation. The first is that at the initial equilibrium

$$k_f \left([H_2O]_0\right) = k_r \left([H^+]_0\right) \left([OH^-]_0\right)$$

The second is that if the perturbation is sufficiently small, ε^2 will be negligibly small. When these two restrictions are incorporated into Eq. (12.23), we get

$$\frac{d\varepsilon}{dt} = - \left(k_f + k_r \left([H^+]_0 + [OH^-]_0\right)\right) \varepsilon \qquad (12.24)$$

We see that this is a first-order equation in ε. Hence the solution is

$$\varepsilon = \varepsilon_0 \exp(-k't) \qquad (12.25a)$$

where k' is the rate constant for change in ε. The equation is usually written as

$$\varepsilon = \varepsilon_0 \exp(-t/\tau) \qquad (12.25b)$$

where τ is called the **relaxation time**. We see from the above equations that

$$\frac{1}{\tau} = k' = k_f + k_r \left([\text{H}^+]_0 + [\text{OH}^-]_0\right) \qquad (12.26)$$

The measured relaxation time for this reaction is 37 μs. The concentrations in the above equation are known, and as the ratio of rate constants is the equilibrium constant $(k_f/k_r = K)$, this is also known. From this information and the above equation we get

$$k_f = 2.5 \times 10^{-5}\text{s}^{-1} \quad \text{and} \quad k_r = 1.4 \times 10^{11}\text{M}^{-1}\text{s}^{-1}$$

Reaction of hydrogen and hydroxyl ions to form water is perhaps the fastest known chemical reaction. The rate is faster than what would normally be expected for a diffusion-controlled reaction between ions in a dielectric medium. This apparent anomaly is explained by the extensive hydrogen bonding in water. Proton transfer in water takes place through intermediate water molecules as indicated in Fig. 12.4. Thus the effective cross-section for H^+ is much larger than that of a single proton and contributes to an enhanced rate.

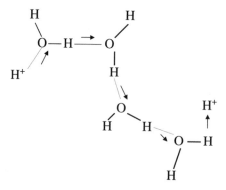

Fig. 12.4.   A mechanism for transfer of protons across the hydrogen-bonded chain of water molecules.

12.5. Comments on Rate Theories

Over a limited variation in temperature, rate laws obey the following equation suggested by Arrhenius in 1889:

$$k = A \exp\left(-\frac{E_a}{k_B T}\right) \tag{12.27}$$

This is one of the most successful equations in chemistry. In this equation A is simply called the pre-exponential factor and E_a, activation energy. The rate constant in the above expression is often a product of rate constants for elementary steps in the overall reaction.

The fraction of molecules with energy E_a or higher increases with temperature, as shown by the exponential term in the above equation. Thus the reaction rate increases with temperature, as this equation predicts.

The idea of activation is also central to **collision theory** and **transition state theory** of chemical reactions. In the former, the rate constant for a bimolecular reaction is equated to the product of the number of collisions and an exponential factor as follows:

$$k = N_A \pi d_{AB}^2 \langle c_{AB} \rangle \exp\left(\frac{-E_a}{k_B T}\right) \tag{12.28}$$

Compare this with Eq. (11.8). The absence of the term $(N/V)^2$ is expected since it appears in the *rate law* expression and not in the *rate constant* expression. The Avogadro number is there to make the units moles instead of molecules. An alternative form for Eq. (12.28) is

$$k = A'\sqrt{T} \exp\left(\frac{-E_a}{k_B T}\right) \tag{12.29}$$

This shows that the results of collision theory differ from the Arrhenius expression by the factor of $\sqrt{T}$. Unless experiments are done over a wide temperature range, the contribution of this factor will be negligible.

The **transition state theory** considers that the two reactants in a bimolecular reaction form an activated complex, which is

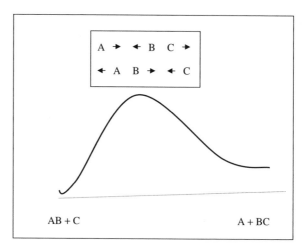

Fig. 12.5. The AB molecule and C form an activated species. The activated species may yield the products or go back to the reactants because of the asymmetric stretch.

unstable to displacement along one normal coordinate. Let us take the reaction

$$AB + C \rightleftharpoons ABC^* \rightarrow A + BC$$

as an example. AB and BC are diatomic molecules; C and A are atoms. ABC^*, the activated complex, has three normal modes, but only the asymmetric stretch can take ABC^* into the products or back into reactants, as shown in Fig. 12.5. According to the transition state theory, this can only happen if ABC^* is meta-stable for the asymmetric stretch, but stable with the other modes of displacement. After some mathematical meandering, the theory gives the following expression for the rate constant:

$$k = \frac{k_B T}{h} \frac{z^*_{ABC}/V}{(z_{AB}/V)(z_c/V)} \exp\left(\frac{-E_a}{k_B T}\right) \qquad (12.30)$$

There is an exponential factor here, as in previous theories. The symbol z denotes the partition function for the species indicated in the subscript. The asterisk on the partition function for the complex indicates that the meta-stable mode is not included in the construction of the partition function. The factor $(k_B T/h)$ has the units of frequency $[s^{-1}]$. This is the frequency for the dissociation of the complex.

What is sketched here is a statistical thermodynamic theory. There is a thermodynamic version of the theory, but we need not visit it to understand the conceptual foundations of chemical kinetics.

12.6. Radiation and Matter

12.6.1. *Spontaneous and stimulated emission*

We will consider just two energy levels of a molecule without degeneracy. Figure 12.6 shows the absorption and emission processes. There are two channels for emission. Spontaneous emission is a random process that is unrelated to radiation in the container except for the frequency of emission. Stimulated emission is a coherent process in which the emitted photon retains the phase, frequency and polarization of the stimulating photon. In 1917, Einstein derived Planck's radiation law from the Boltzmann distribution law using this model. Since my interest is to show the relation between the two types of emission, I will start with these laws.

Planck's law relates energy density per unit frequency, $\rho(\nu)$ $[\text{J m}^{-3} \text{ s}^{-1}]$, to frequency $\nu [s^{-1}]$ of radiation

$$\rho(\nu) = \frac{8\pi h\nu^3}{c^3} \frac{1}{e^{h\nu/k_B T} - 1} \quad (12.31)$$

Here c is the speed of light and h the Planck constant.

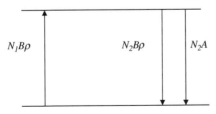

Fig. 12.6. From left to right, absorption, stimulated emission and spontaneous emission. Even though the coefficient, B, and the radiation density, ρ, are the same, the stimulated emission rate is lower than the absorption rate because the upper level has a lower population.

The Boltzmann law for the two-level system is

$$N_1 = N_2 e^{h\nu/k_B T} \qquad (12.32)$$

In a steady state the number of molecules leaving the lower level must be equal to the number returning. Hence we have the following equality for the number of molecules in each level under steady state conditions

$$N_1 B\rho = N_2 B\rho + N_2 A \qquad (12.33)$$

In this equation A is the coefficient for spontaneous emission and B is the coefficient for absorption and stimulated emission. We are assuming that the coefficients for stimulated emission and absorption are equal, rather than explicitly showing it. A photon hitting a molecule has equal probability of being absorbed or stimulating emission. However, the rates of upward and downward *transitions* caused by photons will not be the same since $N_2 < N_1$. Thus we need spontaneous emission to maintain the steady state.

If you substitute Eq. (12.32) into Eq. (12.33) and rearrange to isolate ρ, you will get

$$\rho(\nu) = \left(\frac{A}{B}\right) \frac{1}{e^{h\nu/k_B T} - 1} \qquad (12.34)$$

Comparing this with Planck's law we see that

$$A = \left(\frac{8\pi h\nu^3}{c^3}\right) B \qquad (12.35)$$

This equation shows that spontaneous emission probability increases with frequency. The two probabilities are equal at 69.8 μm or 4.3 THz (border of far infrared and microwaves), if the molecules are in thermal equilibrium at 25°C. Above this frequency spontaneous emission dominates.

For light amplification by stimulated emission (a laser) spontaneous emission is noise to be avoided. How can we then increase the probability of stimulated emission at high frequencies? Detailed analysis shows that if $N_2 > N_1$ (population is inverted) the probability of stimulated emission dominates. However, according to

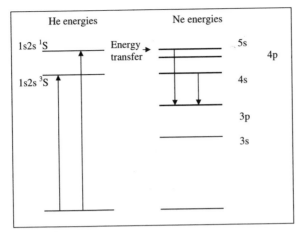

Fig. 12.7. Energy levels involved in helium-neon laser operation. Excitation energy of helium is transferred to excited levels of neon, creating population inversion in neon.

Eq. (12.33) you cannot invert population by increasing temperature, so in actual practice population inversion is done in three or four-level systems.

In a helium-neon laser, helium is excited by electron impact. Excited helium atoms transfer their energy to upper states of neon as shown in Fig 12.7, and effect population inversion in neon. In organic dye lasers, population inversion is between the upper electronic term and higher vibrational levels of the lower electronic term, which are not normally populated. This is illustrated in Fig. 12.8.

12.6.2. *Electronic energy transfer within a molecule*

To illustrate the most significant points, just two energy terms for a polyatomic molecule are shown in Fig. 12.9. Since polyatomic molecules have several vibrational degrees of freedom, we cannot draw a potential energy diagram as we did for diatomic molecules. The best we can do is to indicate by horizontal lines the relevant energy states. Generally speaking, the rotational-vibrational states of the lowest electronic term overlap with similar states in the upper electronic terms. In polyatomic molecules, however, the density

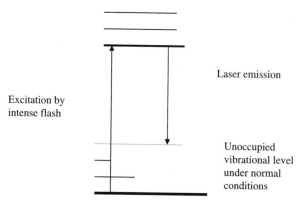

Fig. 12.8. In some dye lasers, excitation is done by flash discharge. Population inversion and laser action occur between the upper electronic term and the higher vibrational states of the lower electronic term. In the figure, the length of the bars of vibrational levels in the lower electronic level indicates that populations in these states fall off exponentially.

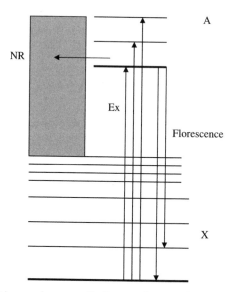

Fig. 12.9. Radiative and non-radiative pathways. Excitation of states in X term result in florescence and transfer of energy to the high energy vibrational-rotational states (shown in the shaded area) of the lower electronic term. The density of states in the shaded area could be very large to form a continuum. This favors non-radiative transfer of energy (NR: non-radiative; Ex: excitation).

of states (number of states per unit energy) of the lower term overlapping with the upper term can be enormous. We already know that the vibrational levels get crowded as energy increases. Beyond this there are several ways in which energy can be distributed when there are many modes. In Fig. 12.9 the overlapping states from the lower term are shown as a continuum, which represents the state of affairs in large molecules.

When the molecule is excited it may not emit at all, or it may emit with a quantum yield (the number of photons emitted for the number absorbed) of less than unity. This is because excitation energy may be transferred to the continuum (non-radiative transfer) and trickle down as low-energy quanta.

If you are interested in measuring emission, non-radiative transitions are a nuisance. But think of what life on this planet would be like if chlorophyll molecules, which absorb solar energy, were to emit with a quantum yield of unity? This illustrates the importance of studying energy transfer through non-radiative processes.

12.6.3. *Electronic energy transfer between two molecules*

The excitation energy of molecule A may return as emission or be transferred to another molecule, B. The kinetic process is shown in Fig. 12.10.

Under constant illumination the rate of production of A^* is equal to the rate of its disappearance. Hence

$$I_a = I_e + k_2[A^*][B]$$

From this we have that

$$\frac{I_a}{I_e} = 1 + \frac{k_2}{I_e}[A^*][B] = 1 + \frac{k_2}{k_1}[B]$$

The last step follows from the equality, $I_e = k_1[A^*]$. We get k_1 by measuring the emission without B and then k_2, the energy transfer rate from the above equation.

Chemical kinetics is a very important and extensive area, and the variety of reactions studied and the resulting techniques that

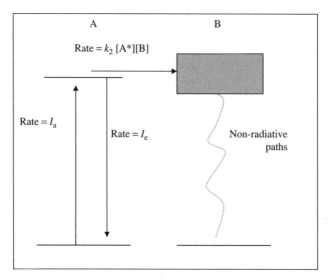

Fig. 12.10. Schematic representation of energy transfer mechanism. Intensities of absorbed (I_a) and emitted (I_e) radiation are the rates of excitation and emission (photons second^{-1}). The rate for energy transfer depends on the concentration of excited A molecules [A*] and [B].

are developed form a large part of chemistry. Yet the fundamental question in all these studies is about mechanism. We can modulate the rate of the overall reaction only when we know its mechanism. In this chapter we have covered the mechanisms for a typical gas phase reaction and a diffusion-controlled reaction in solution. Other significant topics introduced in this chapter are the order of a reaction, half-life and its relation to the order and theories of rates. Common to many rate theories is the existence of a meta-stable activated complex. The interaction of radiation with matter, an important area of science, is described as a kinetic process.

Related material on the disc.

With CD-ROM